The Life Sciences in Early Modern Philosophy

The Life Sciences in Early Modern Philosophy

EDITED BY OHAD NACHTOMY
and
JUSTIN E. H. SMITH

OXFORD
UNIVERSITY PRESS

OXFORD
UNIVERSITY PRESS

Oxford University Press is a department of the University of Oxford.
It furthers the University's objective of excellence in research, scholarship, and education by publishing worldwide.

Oxford New York
Auckland Cape Town Dar es Salaam Hong Kong Karachi
Kuala Lumpur Madrid Melbourne Mexico City Nairobi
New Delhi Shanghai Taipei Toronto

With offices in
Argentina Austria Brazil Chile Czech Republic France Greece
Guatemala Hungary Italy Japan Poland Portugal Singapore
South Korea Switzerland Thailand Turkey Ukraine Vietnam

Published in the United States of America by
Oxford University Press
198 Madison Avenue, New York, NY 10016

Library of Congress Cataloging-in-Publication Data
The life sciences in early modern philosophy / edited by Ohad Nachtomy and Justin E. H. Smith.
pages cm
Includes index.
ISBN 978-0-19-998731-3 (hardback : alk. paper)
1. Biology—Philosophy. 2. Life sciences—Philosophy. 3. Evolution (Biology)
I. Nachtomy, Ohad, editor of compilation.
QH331.L538 2013
570.1—dc23
2013010357

1 3 5 7 9 8 6 4 2
Printed in the United States of America
on acid-free paper

CONTENTS

ACKNOWLEDGMENTS

This book originated in a workshop hosted by the philosophy department at Princeton University. We would like to thank all participants and especially Dan Garber who made the workshop both feasible and pleasant. We would also like to express our deep gratitude to Liat Lavi, who provided crucial assistance and advice throughout the various stages of editing this book. Financial support for this publication was provided by grant 66/08 from the Israel Science Foundation, as well as by a Faculty Research Grant from Fordham University. We would like to thank two anonymous referees for Oxford University Press and Peter Ohlin and Emily Sacharin for supporting the project and following it through with remarkable care and efficiency.

CONTRIBUTORS

Catherine Abou-Nemeh is assistant professor at Victoria University of Wellington, New Zealand. She completed her PhD at Princeton University, where she also received her MA in history. Her dissertation, entitled "Nicolas Hartsoeker's *Système* of Nature: Physics by Conjecture and Optics by Design in Early Modern Europe," sets a detailed, technical history of ideas of the lens-maker and natural philosopher Nicolas Hartsoeker (1656–1727) within late 17th-century Dutch, French, and German intellectual, social, and cultural contexts. She contributed to the volume *Leibniz y las ciencias empíricas. Leibniz and the empirical sciences,* edited by Juan Antonio Nicolás Marín and Sergio Toledo (Editorial Comares, 2011). Her most recent article, "The Natural Philosopher and the Microscope: Nicolas Hartsoeker Unravels Nature's 'Admirable Oeconomy,'" was published in March 2013 in *History of Science.*

Raphaële Andrault, agrégée de philosophie, doctor of philosophy, is a graduate of the ENS Lyon and now holds a post-doctorate in the University of Lyon. Her PhD thesis "La vie et le vivant. Physiologie et métaphysique chez Spinoza et Leibniz" was defended in 2010 (to be published by Champion, 2013). Her recent publications include a critical edition (Nicolas Sténon, *Discours sur l'anatomie du cerveau* (1669), introd. and annotated by Raphaële Andrault, Paris Classiques Garnier, 2009); "Mathématiser l'anatomie: la myologie de Stensen (1667)," *Early Science and Medicine* vol. 15, n° 4–5, 2010: 505–536; and *La raison des corps. Mécanisme et sciences médicales (1664–1720),* Vrin, to be published in 2013.

Andreas Blank specializes in early modern metaphysics and early modern philosophy of the life sciences. He has been visiting lecturer at the Humboldt University of Berlin, visiting fellow at the Center for Philosophy of Science at the University of Pittsburgh, the Cohn Institute for the History and Philosophy of Science at Tel Aviv University, and the Jacques Loeb Center for the Philosophy of the Life Sciences at Ben-Gurion University, Be'er-Sheva, lecturer at the University of Paderborn, and was a visiting professor at the University of Hamburg in the academic year 2011–2012. His publications include *Der logische Aufbau von Leibniz' Metaphysik* [The Logical Structure of Leibniz's Metaphysics] (Berlin and New York: De Gruyter, 2001); *Leibniz: Metaphilosophy and Metaphysics, 1666–1686* (Munich: Philosophia, 2005); and *Biomedical Ontology and the Metaphysics of Composite Substances, 1540–1660* (Munich: Philosophia, 2010); as well as articles in early modern and early analytic philosophy in journals such as *Annals of Science, British Journal for the History of Philosophy, Canadian Journal of Philosophy, Dialogue, Inquiry, Perspectives on Science, Science in Context, The Leibniz Review,* and *Studia Leibnitiana.*

Karen Detlefsen is associate professor of philosophy and education at the University of Pennsylvania. Her recent and forthcoming publications include "Margaret Cavendish and Thomas Hobbes on Freedom, Education, and Women" (*Feminist Interpretations of Hobbes,* forthcoming); "Margaret Cavendish on the Relation Between God and World" (*Philosophy Compass,* 2009); "Reason and Freedom: Margaret Cavendish on the Order and Disorder of Nature" (*Archiv für Geschichte der Philosophie,* 2007); "Explanation and Demonstration in the Haller-Wolff Debate" (*The Problem of Animal Generation in Early Modern Philosophy,* 2006); and "Supernaturalism, Occasionalism, and Preformation in Malebranche" (*Perspectives on Science,* 2003). She is the editor of *Descartes' 'Meditations': A Critical Guide,* to be published in the new series of *Cambridge Critical Guides to Philosophy,* which will include her paper titled "Teleology and Natures in Descartes' Sixth Meditation." She and Andrew Janiak (Duke University) are starting a long-term collaborative project on the natural philosophy of Émilie Du Châtelet, including an edited translation of her *Institutions de physique,* and a monograph on her natural philosophy, especially focusing on her responses to Cartesianism, Newtonianism, and Leibnizianism.

Peter Distelzweig completed his PhD in the History and Philosophy of Science at the University of Pittsburgh and is a visiting assistant professor at Western Michigan University. His dissertation examines the changing relations between mechanical and teleological explanation in physiology in the first half of the 17th century. His research interests center on metaphysical and epistemological issues arising at the intersection of philosophy, science, and medicine in Early Modernity, with a special emphasis on the presence and influence of various strands of late Aristotelianism. He is coeditor of a forthcoming volume on early modern natural philosophy and medicine, to which he also contributes a chapter on mechanics and mechanism in William Harvey. He also works on Aristotle's natural philosophy and philosophy of science. His "The Intersection of Mathematical and Natural Science: The Subordinate Sciences in Aristotle" recently appeared in *Apeiron.*

François Duchesneau is a professor at Université de Montréal, member of the Royal Society of Canada, and associate member of the Institut d'histoire et de philosophie des sciences et des techniques (Paris). Duchesneau's research focuses on the philosophy of biology and the history of science, particularly the history of life sciences from the 17th to the 20th century. In the philosophy of biology, he has dealt with the genesis and structure of theories, the use of models, reductionism and its limits, and various forms of functional explanation. He has published several books, including *Philosophie de la biologie* (1997); *Les Modèles du vivant de Descartes à Leibniz* (1998); and *Leibniz: Le vivant et l'organisme* (2010). Recent edited volumes include *Leibniz selon les Nouveaux Essais sur l'entendement humain* (with Jeremie Griard, 2006). His numerous articles have appeared in journals such as *History and Philosophy of the Life Sciences, Dix-huitième siècle,* and *Perspectives on Science.* He is collaborating with Justin E. H. Smith on a forthcoming critical edition and English translation of the writings of the controversy between Leibniz and Stahl.

Ohad Nachtomy is associate professor at Bar-Ilan University and Fordham University. Among his recent publications are *Possibility, Agency, and Individuality in Leibniz's Metaphysics,* Springer, The New Synthese Historical Library, 2007; "Leibniz on Nested Individuals," *British Journal for the History of Philosophy,* 2007; "Gene expression and the concept of the phenotype," *Studies in History and Philosophy of Science Part C: Studies in History and Philosophy of Biological and Biomedical Sciences,* vol. 38, issue 1, March 2007: 238–254 (with A. Shavit, and Z. Yakhini); It Takes Two to Tango: Phenotyping and Genotying in Genome Wide Association Studies, *Biological Theory* 42:3, 2009 (with Yaron Ramati, Zohar Yakini, and Ayelet Shavit); Leibniz and the Logic of Life, *Studia Leibnitiana,* 2010; *Machines of Nature and Corporeal Substances in Leibniz,* co-edited with Justin E. H. Smith, Springer, The New Synthese Historical Library, 2010; "Leibniz on Artificial and Natural Machines" in *Machines of Nature and Corporeal Substances in Leibniz;* and "A Tale Of Two Thinkers, One Meeting, and Three Degrees Of Infinity: Leibniz And Spinoza in 1675–78," *British Journal for the History of Philosophy,* 2011. He is currently working on a research program on infinity and life in the early modern philosophy, funded by a grant from the Israel Science Foundation.

Brian W. Ogilvie is associate professor of history at the University of Massachusetts, Amherst. He is the author of *The Science of Describing: Natural History in Renaissance Europe,* which examines the origins of modern botany and zoology in the 16th and early 17th centuries. He has held fellowships from the National Science Foundation, the National Endowment for the Humanities, and the Institut d'Etudes Avancées in Paris. His current book project, *Nature's Bible,* examines insects in European art, science, and religion from the Renaissance to the Enlightenment.

Anne-Lise Rey is a lecturer in history of science and epistemology at Université de Lille. Her primary research centers on the concepts of action and perception in Leibniz. She has published many articles on the relation between science and metaphysics in the 17th and 18th centuries in various journals; she has prepared a translation in French and a commentary on the Leibniz-De Volder correspondence; a collection of articles, entitled *Méthode et Histoire. Quelle histoire font les historiens des sciences et des techniques?* forthcoming in Garnier, 2014. She has co-edited an issue of *Revue de Métaphysique et de Morale* (2011), entitled "Nicolas de Cues et G.W. Leibniz: Infini, expression et singularité," and is preparing an issue for the *Revue d'Histoire des Sciences*, entitled "La guerre en lettres." She is also co-leading the project ANR Philomed, which focuses on medical discoveries and philosophy of human nature.

Lea F. Schweitz is assistant professor of systematic theology/religion and science at the Lutheran School of Theology in Chicago and the director of the Zygon Center for Religion and Science. Her interdisciplinary research occurs at the intersections of early modern history of philosophy, history of the life sciences, and history of theology. She has published in the *Journal of Religion*, *Zygon: Journal of Religion and Science*, and the *Leibniz Review*. She is working on a series of articles on the emerging concepts of humanity in the early modern period, including an essay entitled "Reasoning Faithfully: Leibniz on Reason's Triumph of Faith and Love" to appear in *When Reason Is a Whore: The Quandary of Lutheran Philosophy* (2011). She is also working on a book-length project that explores some of the theological implications for the evolving understandings of nature in the early modern period.

Justin E. H. Smith is university professor of the history and philosophy of science at the University of Paris 7-Denis Diderot. He is the author of *Divine Machines: Leibniz and the Sciences of Life* (Princeton University Press, 2011), and of the forthcoming *Nature, Human Nature, and Human Difference: Early Modern Philosophy and the Concept of Race* (Princeton University Press). With François Duchesneau, he is the editor and translator of *The Leibniz-Stahl Controversy* in the Yale Leibniz Series.

Thomas Teufel is associate professor of philosophy at Baruch College, CUNY. He received his PhD in philosophy from Harvard University in 2006. His primary research interest is in the philosophy of Immanuel Kant and, in particular, Kant's third and final 'Critique,' the *Critique of the Power of Judgment.* His recent publications include "What does Kant mean by 'Power of Judgment' in his Critique of the Power of Judgment?" *Kantian Review*, vol. 17.2 (2012); "What is the Problem of Teleology in Kant's Critique of the Teleological Power of Judgment?" *SATS: Northern European Journal of Philosophy*, vol.

12.2 (December 2011): 198–236; "Kant's Non-Teleological Conception of Purposiveness," *Kant-Studien*, vol. 102.2 (June 2011): 232–252; and "Wholes that Cause their Parts: Organic Self-Reproduction and the Reality of Biological Teleology," *Studies in History and Philosophy of Science, Part C: Studies in History and Philosophy of Biology and Biomedical Sciences*, vol. 42 (June 2011): 252–260.

Charles T. Wolfe is research fellow, Centre for History of Science, Ghent University, and an associate member of the Unit for History and Philosophy of Science, University of Sydney. His work focuses on the interrelation between early modern philosophy and the history and philosophy of the life sciences—primarily medicine, biology, and natural history—centering on themes such as the man-machine, mechanism and organism, vitalism and materialism; and figures such as Locke, La Mettrie, and Diderot. He has published in journals such as *Early Science and Medicine, Perspectives on Science, Bulletin de la SHESVIE, Recherches sur Diderot et l'Encyclopédie*, and *Dix-huitième siècle*; his edited volumes include *Monsters and Philosophy* (2005), a special issue of *Science in Context* on *Vitalism without Metaphysics?* (2008); *The body as object and instrument of knowledge. Embodied empiricism in early modern science* (with Ofer Gal, 2010); *The Concept of Organism* (with Philippe Huneman, special issue of *History and Philosophy of the Life Sciences*, 2010); and *Vitalism and the scientific image, 1800–2010* (with Sebastian Normandin, forthcoming). His current project is a monograph on the conceptual foundations of vitalism.

The Life Sciences in Early Modern Philosophy

Introduction

OHAD NACHTOMY AND JUSTIN E. H. SMITH, EDITORS

When philosophers and historians think of the conceptual innovations in early modern science that played an important role in the shaping of the modern world, it is generally mechanical physics that first comes to mind: the newly mathematized study of the motion of projectiles, planets, and billiard balls. The reason for this focus is that there were real, incontestable advances in the understanding of how the physical world works, and this improved understanding had plainly tangible effects on the history of technology and applied science. However, it is a rather narrow understanding of history that concerns itself only with those domains of human endeavor that are enjoying a phase of rapid progress. For indeed, very often it is precisely the domains of inquiry that are lagging behind that also present the most difficult and interesting conceptual problems, and that weigh most heavily on the minds of the thinkers and experimenters who are simultaneously enjoying the fruits of their progress elsewhere. If in early modern Europe planetary motion was coming to be seen as more tractable than animal motion, this does not at all mean that planets were held to be in more urgent need of explanation than animals. Quite the contrary, one might say that what happened was this: The planets, as it turned out, lent themselves more easily to explanation. They were, one might provocatively suggest, the easier part of the scientific revolution. The harder part, the part that would not emerge as a full-fledged, independent domain of science until the 19th century, concerned that great part of nature that appears resistant to explanation in the same terms as planets and billiard balls: the part of nature that does things that billiard balls do not do; the part that replicates itself, producing nearly identical copies of itself in many different ways; the part that has the capacity for self-motion; the part that eats and breathes.

Now there was, in the early modern period, a great diversity of views as to what exactly the relationship was between this latter domain of nature and the one described by mechanical physics. Arguably the most significant attempt to deal with the gap between these domains (living versus nonliving) was to assimilate the one into the other, that is, to understand the living world in mechanical terms. Many philosophers (a label that includes what are today called 'scientists'), such as, notably, René Descartes, supposed that the success of their own programs of mechanical physics depended upon their ability to explain the generation, structure, and motion of living beings, as Descartes put it, "in the same manner as the rest,"[1] which is to say that Descartes hoped to make living nature comprehensible by appealing to the same principles and laws that were lately helping to make so much sense of the rest of nature. Others, such as the 'vitalist' philosopher Henry More, supposed that the motion and generation of living beings could not possibly be explained "in the same manner as the rest." But the key thing to observe here is that, notwithstanding their opposing commitments, all agreed that it is of the utmost urgency to account for the generation, structure, and motion of living beings. Indeed, Descartes would even assert that the success of his *entire* philosophy depended on his ability to provide a successful account of certain vital processes.[2] And though he does not say as much, we may rightly characterize More's 'entire philosophy' as being tied up with the success or failure of his conviction that, *pace* Descartes, vital processes *cannot* be explained in the way Descartes hoped to explain them.

In sum, no matter which particular views early modern philosophers held about living beings, all agreed on the urgency of providing an account of them. Such an account could either be directed at explaining away their supposed uniqueness or, to the contrary, at proving their supposed uniqueness; either way, an account was in order. In this respect, one might propose that the problem of life in the early modern period imposed itself in much the same way that the famous problem of consciousness does today. David Chalmers and his opponents strongly disagree on the question of whether there *is* a 'hard problem' of consciousness or not,[3] but both sides in the debate spend a great deal of time and energy discussing consciousness. If we follow this analogy through, life ends up being tractable in much the same way Chalmers's opponents suppose consciousness will end up being tractable (i.e., as a natural phenomenon among others, as a result of the arrangement of natural 'stuff'). Living beings turn out to be unusually complicated arrangements of proteins, but their unusual complexity does not, in the end, place them in a distinct ontological class.

[1] AT VI 45. AT: *Oeuvres de Descartes,* 12 vols., eds. C. Adam and P. Tannery (Paris: Vrin/CNRS, 1964–76) (cited by volume and page).

[2] Thus, for example, he wrote in a 1639 letter to Marin Mersenne: "I am prepared to admit that if what I have written on [cardiac motion]... turns out to be false, then the rest of my philosophy is entirely worthless" (AT II 501).

[3] See in particular David Chalmers, "Facing Up to the Problem of Consciousness," *Journal of Consciousness Studies* 2, 3 (1995): 100–219. For a characteristic statement of the opposing view, see Daniel Dennett, "Explaining the 'Magic' of Consciousness," *Journal of Cultural and Evolutionary Psychology* 1, 1 (2003): 7–19.

Things could of course turn out differently for the problem of consciousness; there is no prima facie reason why we should expect it to be affected by science in the same way as life came to be over the course of the 19th and 20th centuries. But the comparison of the two problems, as they occur in history, with each perceived as being 'hot' in its own period and context, may help us to understand what was at stake in early modern reflections on the concept of life and the nature of living beings. For in each case, what we see is disagreement as to whether the phenomenon or entity in question is one for which the simple accumulation of more empirical or scientific information about it will prove to be enough to fully comprehend it. One camp sees the resolution of the problem as being—they hope—dependent simply on accumulation of more knowledge of the sort we already have, while the other camp supposes that the phenomenon or entity in question—life, as the case may be, or consciousness—requires knowledge of a different sort than the kind that is already helping us to make sense of the rest of the world. This latter camp supposes that there is a conceptual difference between living beings and nonliving ones, or between conscious beings and non-conscious ones. And this commitment, in turn, also requires their opponents to engage with the problem at the conceptual level, if only to refute the view that there is a conceptual distinction at all. And so, in the end, all parties to the debate about life or consciousness are engaged in the practice of philosophy.

What those who adhere to the conceptual distinctness of the domain of living beings tend to highlight are a few remarkable features of these beings that seem entirely resistant to explanation "in the same manner as the rest." For one thing, these beings seem to be *structured* entirely unlike anything else in nature. They have complex interworking parts (as, admittedly, do certain other natural but plainly inanimate systems), and moreover this functional complexity seems to operate at very fine-grained levels of analysis. Indeed, prior to the discovery of the living cell in the 19th century, and to some extent even up until the DNA sequencing of a number of biological species, it could easily appear that there simply was no lower limit to vital organization at all (i.e., that the functional complexity would continue beyond any *possible* level of analysis). In this respect, living beings seemed to be in a class entirely apart from other natural entities, which, it was supposed, could be analyzed down to their basic, nonfunctional constituents in a small number of steps.

A second apparent difference between living beings and nonliving entities is that the former seem to operate according to special laws that have no equivalent in the rest of the natural world. For example, if a nerve is pricked, and the limb of which it is a part twitches, this twitch cannot be understood in any obvious way according to the same principles that explain the communication of motion from one billiard ball to another. Now of course those who deny that life constitutes a distinct ontological category can argue that, at the microscopic level, what is going on is fundamentally no different than what goes on with billiard balls, that muscle contraction and expansion involve countless, sub-visible billiard-ball-like collisions. But the onus is on them to find these collisions and to describe them, and in the absence of any plausible means of providing such a description, those who defended the ontological distinctness of living beings not surprisingly were not impressed by the evident explanatory stretching required to extend the language of mechanical physics to the domain of physiology.

Similar points may be made, of course, about other peculiar physiological phenomena, such as digestion, respiration, circulation, etc. But one phenomenon has always stood out as a particularly salient motivation for claiming that living beings constitute a distinct ontological class: namely, their capacity for reproduction, for cycling back upon themselves, as Aristotle would put it, thus obtaining a sort of eternity in number if not in kind.[4] Now today we are starting to see some rudimentary examples of machine self-replication, but in the 17th century, it seemed a fairly solid and certain shibboleth for the separation of the living from the nonliving, or of beings from machines and artifacts, that whereas no two clocks placed in a room together ever managed to produce a third, little clock, two dogs or humans, if appropriately selected, very well might.[5] Asexual generation was already well attested (though differently described), and spontaneous generation was still debated; these do not involve the coming together of two individuals for the production of a third, but they do seem to be processes that have no analog in nonliving nature. Spontaneous and asexual generation, along with sexual generation, all seem to yield up *beings* in the strict sense, rather than simply yielding new arrangements of preexisting matter, as would seem to be the case in, say, the formation of stalactites or (a more controversial, indeed borderline, case) crystals. This is to say that living beings, in contrast to natural entities, are the sort of things that are *generated* rather than simply formed, and what generation is, exactly, and how it is that a being that previously did not exist, later does, is not just a philosophical problem, but at least since the pre-Socratics has stood as a philosophical problem par excellence.

[4] Aristotle writes of animals in *On the Generation of Animals* II 1: "For since it is impossible that such a class of things as animals should be of an eternal nature, therefore that which comes into being is eternal in the only way possible. Now it is impossible for it to be eternal as an individual (though of course the real essence of things is in the individual)—were it such it would be eternal—but it is possible for it as a species. This is why there is always a class of men and animals and plants."

[5] See in particular this passage from a 1683 letter by Bernard de Fontenelle: "Do you say the beasts are machines just as watches are? Put a dog machine and a bitch machine side by side, and eventually a third little machine will be the result, whereas two watches will lie side by side their whole lives without ever producing a third watch" (The original passage can be found in *Oeuvres de Fontenelle,* des Académies Française, des Sciences, des Belles-lettres, de Londres, de Nancy, de Berlin et de Rome, vol. 5, 99.)

One further way in which living nature stands apart—and also a way that, in turn, stands apart from those already listed—is in respect of its order and variety, and the nature of the kinds of living things that make up this variety. Of course, there are chemical and physical kinds too, but as many early modern philosophers insisted, it is living kinds, what we now call biological species, that seem to represent the very idea of what would later be called, after J. S. Mill, 'natural kinds.'[6] Biological species, in the early modern period as today, seem to play an important role in constituting our idea of what it is for a thing to be a thing of a certain sort: A moment's reflection will convince you that the great majority of examples typically adduced by philosophers to illustrate the idea of form, essence, haecceity, and so on, have always been biological species. This means, inevitably, that thinking about the living world entered in important ways into classical philosophical debates, such as that between the nominalists and the realists, and it would be neglectful to gloss over the way in which biological species imposed their particular character here, or to suppose that the examples of, say, 'cowhood' and 'triangularity' always served the same conceptual purposes, in the same way, in the course of philosophical arguments about whether there are real kinds or not. In the early modern period in particular, the longstanding practice of allowing biological species to stand in as paradigmatic examples of kinds in general was thrown into a sort of crisis, as the rate of discovery of new kinds sky-rocketed as a consequence of microscopic investigations, as a result of greatly increased contact with the non-European world. Empirical discoveries greatly complexified the understanding of natural order, and indeed called into question the very idea that nature constitutes an order at all.

The late 17th and early 18th centuries were a period of intense contestation of the boundaries between different domains of nature. While the most familiar narrative of this period tells us that it was a time of "mechanization," in which nature came to be considered as a system of interworking mechanical parts, in fact no sooner was the mechanical hypothesis put forward than it faced intense criticism and opposition. Much of the most significant opposition came from thinkers who were concerned with the phenomena of life, and many of these, such as Leibniz, may be described as attempting to reenchant mechanical nature by positing active or mind-like principles as necessary for the explanation of even the most basic and universal natural phenomena, particularly motion. The problem of life, in this respect, is at the heart of some of the most fundamental debates in early modern natural philosophy and metaphysics.

We have organized the contributions to this volume in accordance with the particular problems, just outlined, that living beings and living nature posed for early modern philosophy: the problem of life in general, whether it constitutes

[6] See in particular J. S. Mill, *A System of Logic*, London: Longman, 1884.

something ontologically distinct at all, or whether it can ultimately be exhaustively comprehended "in the same manner as the rest"; the problem of the structure of living beings, by which we understand not just bare anatomy but also physiological processes such as irritability, motion, digestion, and so on; the problem of generation, which might be included alongside digestion and other vital processes were it not for the fact that it presented such an exceptional riddle to philosophers since antiquity, namely, the riddle of coming-into-being out of—apparent or real—nonbeing; and, finally, the problem of natural order.

The scholarly investigation of the intersection between philosophy and the life sciences in the early modern period is not only interesting in its own right; it is also helping to unravel many well-entrenched historiographical and interpretive categories, such as the distinction between rationalists and empiricists and the distinction between realists and idealists. These paradigmatic distinctions now appear rather less salient for apprehending what was actually at stake in early modern philosophical debates. In their place, and in part through the work of the scholars collected here, new approaches are being developed. The present volume advances a recent historiographical turn toward the intersection of early modern philosophy and the life sciences by bringing together many of its leading scholars to present the contributions of important but often neglected figures, such as Ralph Cudworth, Nehemiah Grew, Francis Glisson, Hieronymus Fabricius ab Aquapendente, Georg Ernst Stahl, Juan Gallego de la Serna, Nicholas Hartsoeker, Henry More, as well as more familiar figures, such as Descartes, Spinoza, Leibniz, Malebranche, and Kant.

Throughout, the contributors to the volume are concerned, each in their own way, with revealing the crucial importance for early modern philosophy of questions in what would later branch off as the science of biology. They are concerned with showing not that some independent area of activity conceptualized as the 'life sciences' was of interest to early modern philosophers, in the way that today, say, biology is of interest to philosophers of biology, but rather that what we today think of as questions of the life sciences were in fact *constitutive* of what early modern philosophers themselves thought of as philosophy. This is, or ought to be, beyond any historical doubt. We hope that the essays in this volume will help bring our scholarly community—particularly historians of philosophy and philosophers with a serious interest in history—somewhat closer to where we ought to be.

PART ONE

THE NATURE OF LIVING BEINGS

1

Infinity and Life

The Role of Infinity in Leibniz's Theory of Living Beings

OHAD NACHTOMY

1. Introduction: What Does Infinity Have to Do with Life?

What does infinity have to do with life and the nature of living beings? At first glance, the answer seems to be "not very much." Infinity seems to belong primarily to mathematics and, since human observational capacities are finite and limited, infinity (or any infinite thing, for that matter) cannot be observed in nature. Thus it would seem that infinity would have no place in an adequate description of nature, in general, and in its living part, in particular.

And yet infinity does play a major role in some important views of the natural world. One need only recall that one of the most crucial transformations from the old to the new science was described as a turn from a closed world to an infinite universe.[1] The story of the role this transformation played in the scientific revolution (in the 16th and 17th centuries), as well as its impact on human psychology and imagination, is quite well known (if perhaps somewhat

[1] A. Koyré, *From the Closed World to the Infinite Universe* (Baltimore, MD: Johns Hopkins University Press, 1957). Abbreviations are as follows: A: G. W. Leibniz, *Sämtliche Schriften und Briefe*, Darmstadt/Leipzig/Berlin: Edition of the German Academy of Sciences, 1923–, cited by series, volume, and page. If not otherwise indicated, the reference is to series 6, vol. 3. AG: G. W. Leibniz, *Philosophical Essays*, D. Garber and R. Ariew (eds. and trans.) Indianapolis,IN: Hackett, 1989. GP: *Die Philosophischen Schriften von G. W. Leibniz*, C. I. Gerhardt (ed.), 7 vols. Berlin: Weidmann, 1875–90; reprinted Hildesheim, DE: Olms, 1978. GM: *Die mathematischen Schriften von G. W. Leibniz*, C. I. Gerhardt (ed.), Berlin: Winter, 1860–75; LLC: G. W. Leibniz, *The Labyrinth of the Continuum. Writings on the Continuum Problem, 1672–1686*, translated by R. Arthur, New Haven, CT, and London, Yale University Press, 2001. *Theodicy*: G. W. Leibniz, *Essays on the Goodness of God the Freedom of Man and the Origin Evil*, translated by E. M. Huggard (La Salle, IL: Open Court, 1993), first published (London: Routledge & Kegan Paul, 1951).

underappreciated). At the same time, it is much less known and much less appreciated that the turn from a finite and closed world to an infinite one has taken place not only in cosmology and astronomy but also in the life sciences. In other words, the change from the finite to the infinite has taken shape not only at the macroscopic level, regarding the non-limited nature of space, but also at the microscopic level where worlds within worlds of minute animals (*animalcula*) have been discovered or postulated.

One might say that what the invention of the telescope has given the scientific imagination at the macroscopic level, the invention of the microscope has given the scientific imagination at the microscopic level—that is, some empirical insight into new phenomena and a lot to stimulate the scientific imagination.[2]

In fact, during this time period scholars were preoccupied not only with the infinitely large but also with the infinitely small. Thus, infinitesimals were used not only in pure mathematics but also in describing and understanding motion (as exemplified in Hobbes's notion of conatus); forces (as exemplified both in Leibniz's and Newton's work); the division of matter; and, what is the main concern of this project, the nature of living beings.[3] The work of early microscope observers, such as Malpighi and Leeuwenhoek, is most pertinent here. Pascal's example of a mite stands out as an illustration of this preoccupation. Pascal's eloquent remarks regarding the infinitely large and the infinitely small may serve here as a reminder of the role infinity plays in both domains. Pascal writes:

> What is a man in the infinite? Who can comprehend it? But to show him another prodigy equally astonishing, let him examine the most delicate things he knows. Let a mite be given him, with its minute body and parts incomparably more minute, limbs with their joints, veins in the limbs, blood in the veins, humours in the blood, drops in the humours, vapours in the drops. Dividing these last things again, let him exhaust his powers and his conceptions, and let the last object at which he can arrive be now that of our discourse. Perhaps he will think that here is the smallest point in nature. I will let him see therein a new abyss. I will paint for him not only the visible universe, but also everything he is capable of conceiving of nature's immensity in the womb of this

[2] It is an open question how important a role these devices play in these transitions, but there is no doubt that both play a considerable role in the development of astronomy, as well as the life sciences. Leibniz himself noted that one could have anticipated the existence of microscopic animals prior to their empirical discovery just "as Democritus foresaw the imperceptible stars in the Milky Way before the discovery of the telescope." See footnote 23 for the reference and some discussion of this point.

[3] It is also worth recalling Giordano Bruno's lines in this context: "The minimum of nature or reality is amazingly smaller than the smallest perceptible minimum. There is no art to define it." De triplici minimo et mensura, LIB.i cap. 9.

> imperceptible atom. Let him see therein an infinity of worlds, each of which has its firmament, its planets, its earth, in the same proportion as in the visible world; in this earth of animals, and ultimately of mites, in which he will find again all that the first had, finding still in these others the same thing without end and without cessation. Let him lose himself in wonders as amazing in their littleness as the others in their vastness.[4]

While the mathematical, astronomical, and physical facets of the scientific revolution have been studied in great detail, the history of the life sciences and its development during the 16th and 17th centuries deserve more attention. Specifically, the role infinity plays in the early stages of the life sciences and the way it figures in various conceptualizations of living beings certainly deserve more attention.

Part of the reason for this lack of attention may be due to the way infinity was considered by some major figures in the making of the scientific revolution. Both Descartes and Pascal, who were among the most important mathematicians of the time, were skeptical about using infinity in a scientific description of nature. They argued that, due to the finite nature of our mind, the infinite is incomprehensible to us and must remain beyond our grasp. In stressing the contrast between the finite and the infinite, Pascal and Descartes delineated an irreconcilable gap between the infinite creator and its finite creatures, suggesting

[4] The passages continues thus: "For who will not be astounded at the fact that our body, which a little while ago was imperceptible in the universe, itself imperceptible in the bosom of the whole, is now a colossus, a world, or rather a whole, in respect of the final smallness which we cannot reach? He who regards himself in this light will be afraid of himself, and observing himself suspended in the mass given him by nature between those two abysses of the Infinite and Nothing, of which he is equally removed, will tremble at the sight of these marvels; and I think that, as his curiosity changes into admiration, he will be more disposed to contemplate them in silence than to examine them with presumption." Blaise Pascal, *Pensées de M. Pascal sur la religion et sur quelques autres sujets* (Paris, 1670), XXII, 169–175; the English translation is from Lloyd Strickland's website (http://www.leibniz-translations.com/pascal.htm#note2).

Leibniz's response to Pascal's remarks (in a note he later deleted) is extremely interesting: The full English translation reads: "What he has just said on the subject of the double infinity is only an entrance into my system. What would he not have said with that power of eloquence he possessed if he had gone further, if he had known that all matter is organic, and that the least portion contains, through the actual infinity of its parts in an infinity of ways, a living mirror expressing the whole infinite universe, so that one could read in it (if one had sufficiently keen sight and mind) not only the present stretched out to infinity, but also the past, and the whole infinitely infinite future, since it is infinite in every moment, and there is an infinity of moments in each part of time, and more infinity than one could express in the whole of future eternity." *G. W. Leibniz: Textes inédits d'après les manuscrits de la Bibliothèque provinciale de Hanovre*, G. Grua (ed.), Paris, 1948, 553; F. de Buzon 554. The English translation is from Lloyd Strickland's website (http://www.leibniz-translations.com/pascal.htm#note2).

that it would be not only cognitively impossible but also morally and theologically misguided for us to investigate the infinite.[5]

This surely contributed to a hostile attitude for the use of infinity beyond its theological context (as an essential attribute of God). Galileo, for his part, exposed the paradoxical nature of the infinite, arguing that even the most basic relations of the quantitative sciences, such as 'larger than', 'smaller than', or 'equal to', cannot be applied when we are dealing with the infinite.[6] If one accepts Galileo's reasoning, it would follow that infinity cannot be used in a scientific description of nature for, according to the norms of the new philosophers, such a description must surely involve quantification.

2. Infinity in Leibniz

At the same time, when we turn our attention to Leibniz, we see that infinity figures in almost every aspect of his philosophy. Infinity plays a crucial role not only in his views of God, space, number, and possible worlds, but also in his views of the natural world, as can be seen in his response to Pascal.[7] Leibniz is not wary of using infinity. Rather, he believes that infinity should be admired and investigated precisely because created things, which he identifies with living beings, bear the mark of their author, that is, a mark of perfection and infinity imprinted in their very nature.

In his early work on the infinitesimal calculus (initiated in Paris under the guidance of Huygens), Leibniz discovered a rational method to treat infinity in mathematics. By translating infinitesimal quantities into finite ones, arguing that they can be regarded as variables, smaller or larger than any assignable quantity, he showed that infinitesimals could in fact be used in calculations.[8] Leibniz's sophisticated approach (evident in his early work in mathematics) certainly contributed to his applying infinity in other domains of his philosophy as well.

[5] Descartes's *Principles of Philosophy*, Part I, articles 26–27. See Pascal's *Pensées* fragments 194–427, 104–105 in Galimard–Flamarion edition (1976). As Pascal notes, we perceive the infinite but do not understand its nature (fragments 233–418, Galimard–Flamarion edition (1976), 113).

[6] In his *Discourses and Mathematical Demonstrations Concerning the Two New Sciences,* Galileo writes: "For I believe that these attributes of greatness, smallness, and equality do not befit infinities, about which it cannot be said that one is greater than, smaller than, or equal to one another" (Edition Nazionale 77–78; cited from Richard Arthur's translation in LLC 355).

[7] It is also worth mentioning the role infinity plays in the resolution of the two "labyrinths" the human mind gets entangled in: "the labyrinth of the continuum" and "the labyrinth of human freedom" (AG 95).

[8] See Arthur's introduction to LLC and his "Actual Infinitesimals in Leibniz's Early Thought" in *The Philosophy of the Young Leibniz,* Studia Leibnitiana Sonderheft, edited by M. Kulstad, M. Laerke, and D. Snyder (Stuttgart, DE: Franz Steiner Verlag, 2009), 11–28.

For, given this approach, one could feel free using infinity without falling into paradox.[9]

And Leibniz indeed felt free to use infinity in his philosophy. Thus, for example, he held that the actual world is one of infinitely many possible worlds and that possible worlds, in turn, are conceived by God's infinite intellect[10] (which is the region of all possibilities). God himself is seen as an *infinite* and most perfect being. The actual world, too, consists of infinitely many individual substances, each of which involves relations to infinitely many others and "exhibits an infinite series of operations" (Comments on Fardella, AG 102). Likewise, for Leibniz, "[e]ach portion of matter may be conceived as a garden full of plants and as a pond full of fishes. But each branch of every plant, each member of every animal, each drop of its liquids is also some such garden or pond" (*Monadology* §67; AG 222).

Leibniz clearly applies infinity both to the large and to the small. Far from avoiding the infinite in his metaphysics, as Descartes and Pascal recommend, Leibniz is celebrating infinity. He obviously finds it essential for an adequate description of the world.[11] As he writes to Foucher:

> I am so much in favor of actual infinity that, instead of admitting that nature rejects it, as it is vulgarly said, I hold that it affects it everywhere, *for better marking the perfections of its author*. Thus I believe that there is no part of matter which is not, I do not say divisible but actually divided and, by consequent, the least particle would be considered as a world full of an infinity of diverse creatures. (Letter to Foucher, GP I 416)[12]

[9] Leibniz's approach is far more complicated. He certainly rejects infinite number (and infinite quantities) as paradoxical but shows that the syncategorematic sense of infinity avoids these paradoxes. In addition, he distinguishes between different kinds and different degrees of infinity.

[10] *Monadology* §53; *Theodicy* §225.

[11] «Mes méditations fondamentales roulent sur deux choses, savoir sur l'unité et sur l'infini. Les âmes sont des unités et les corps sont des multiplicités, mais infinies tellement que le moindre grain de poussière contient un monde d'un infinités des créatures » (Letter to Sophie, November 4, 1696).

[12] «Je suis tellement pour l'infini actuel, qu'au lieu d'admettre que la nature l'abhorre, comme l'on dit vulgairement, je tiens qu'elle l'affecte partout, pour mieux marquer les perfections de son auteur. Ainsi je crois qu'il n'y a aucune partie de la matière qui ne soit, je ne dit pas divisible, mais actuellement divisée, et par conséquent, la moindre particelle doit être considérée comme un monde plein d'une infinité des créatures différentes» (Letter to Foucher, GP I 416). See also Monadology §65: "Every portion of matter is not only divisible to infinity, as the ancients realized, but is actually subdivided without end, every part into smaller parts, each part divided into parts having some motion of their own" (AG 221). Compare this to Aristotle's view expressed in his *Generation of Animals*: "Nature avoids what is infinite, because the infinity lacks completion and finality, whereas this is what nature always seeks" (1.1.715b15). As I argue below, there is a sense of infinity in Leibniz that means precisely absolute perfection and completion. But, of course, this is not completion in Aristotle's teleological sense.

While this passage makes clear that Leibniz is keen on using infinity in his description of the natural world, I would like to stress that Leibniz employs different notions of infinity in different contexts. He is particularly careful to distinguish between using infinity in mathematical contexts, which usually pertain to abstract and ideal entities, and in metaphysical contexts, which usually pertain to concrete and real beings. This approach is exemplified by Leibniz's position regarding the labyrinth of the continuum. According to Leibniz, the labyrinth of the continuum is produced when we confuse the mathematical and the ideal realms, which admits of potential infinity, and the metaphysical or the real, which admits of actual infinity. Likewise, the way out of the labyrinth would be to disentangle the confusion by carefully distinguishing between the two realms and the corresponding types of infinity.[13]

I shall return to this point in attempting to discern the exact sense in which Leibniz ascribes infinity to living beings. But before we come to focus on the different ways Leibniz employs infinity, I would like to substantiate and enrich the claim that Leibniz is using infinity in describing (and in defining) living beings by reviewing some relevant texts set in a chronological order.

2.1 Infinity and Life in Leibniz

As early as his Theory of Concrete Motion (1670–71) Leibniz articulates the doctrine that there are worlds within worlds to infinity. In this context the doctrine appears as a consequence of the infinite divisibility of the continuum. As he writes, "any atom will be of infinite species, like a sort of world, and there will be worlds within worlds to infinity" (A 6.2 N40; LLC 338–339). A similar view appears in Leibniz's notes from Paris (1676), where he writes that every part of the world, regardless of how small, "contains an infinity of creatures" which is itself a kind of "world" (A 6.3 474).[14]

In a letter to Malebranche, from 1679, Leibniz writes:

> There is even room to fear that there are no elements at all, everything being effectively divided to infinity in organic bodies. For if these microscopic animals are in turn composed of animals or plants or other

[13] "As long as we seek actual parts in the order of possibles and indeterminate parts in aggregates of actual things, we confuse the ideal things with real substances and entangle ourselves in the labyrinth of the continuum and inexplicable contradictions" (Leibniz letter to De Volder, January 19, 1706, translated in AG 185). See also Pauline Phemister, *Leibniz and the Natural World Activity, Passivity and Corporeal Substances in Leibniz's Philosophy* (Dordrecht, DE: Springer, 2005) and Richard Arthur's introduction to LLC.

[14] G. W. Leibniz, *De Summa Rerum: Metaphysical Papers 1675–76*, edited and translated by G. H. R. Parkinson (New Haven and London: Yale University Press, 1992) 25.

> heterogeneous bodies, and so on to infinity, it is apparent, that there would not be any elements. (A Iii 719, translation in Smith 2010, 235)

In *Primary Truths* (1689–90), Leibniz again notes that "every particle of the universe contains a world of an infinity of creatures" (A 6.4 1647–48).[15]

In the course of his correspondence with Arnauld, it begins to become clear that what Leibniz has in mind by these creatures (that are to be found everywhere) is living things:

> [A]nd since matter is infinitely divisible, no portion can be designated so small that it does not contain animated bodies, or at least bodies endowed with a primitive Entelechy or (if you permit me to use the concept of life so generally), with a vital principle; in short, corporeal substances, of all of which one can say in general that they are living. (G II 118)

Furthermore, in a letter (to Arnauld) from November 28, 1686, Leibniz suggests that living (or being animate) is the mark of a true corporeal substance:

> I cannot say precisely whether there are true corporeal substances other than those that are animated, but souls at least serve to give us some knowledge of others by analogy. (AG 79)

In the same letter, Leibniz clarifies that being animate is the only limit to the infinite divisibility of matter:

> For the continuum is not merely divided to infinity, but every part of matter is actually divided into other parts as different among themselves as the aforementioned diamonds. And since we can always go on in this way, we could never reach anything about which we could say, here is truly a being, unless we found *animated machines* whose soul or substantial form produced a substantial unity independent of the external union arising from contact. (My italics, AG 80)

As we can see in these passages, in the correspondence with Arnauld, Leibniz begins to use the notion of life as a criterion for being, and he is clearly ascribing

[15] In a note on a letter of Michelangelo Fardella from 1690, we find again Leibniz's commitment to the worlds within worlds ad infinitum. He writes: "... there are substances everywhere in matter, just as points are everywhere in a line ... [J]ust as there is no portion of a line in which there is not an infinite number of points, there is no portion of matter which does not contain an infinite number of substances" (AG 105).

infinity to living beings, so that they are to be found everywhere (or in the least part of matter).[16]

From 1695, beginning with the *New System of Nature,* Leibniz is no longer using infinity merely in a general way (of describing nature as worlds within worlds to infinity). From now on, he is clearly using infinity as one of the defining features of living beings. In arguing (in the *New System*) against Descartes's claim that living things are nothing but subtle machines, akin to artificial machines, though more subtle and complex, Leibniz argues that natural machines are essentially different than artificial machines in that they are machines to the least of their parts, so that they are machines within machines ad infinitum.

> I believe that this conception (in which the difference between natural machines and ours is merely one of degree) does not give us sufficiently just and worthy idea of nature, and that my system alone allows us to understand the true and immense distance between the least production and mechanisms of divine wisdom and the greatest masterpieces that derive from the craft of a limited mind; this difference is not simply a difference of degree, but a difference in kind. We must then know that the machines of nature have a truly infinite number of organs, and are so well supplied and so resistant to all accidents that it is not possible to destroy them. A natural machine still remains a machine in its least parts, and moreover, it always remains the same machine that it has been, being merely transformed through different enfolding it undergoes, sometimes extended, sometimes compressed and concentrated as it were, where it is thought to have perished. (*New System of Nature,* AG 142)
>
> Moreover, a natural machine has the great advantage over an artificial machine, that, displaying the mark of an infinite creator, it is made up of an infinity of entangled organs. And thus, a natural machine can never be absolutely destroyed just as it can never absolutely begin, but it only decreases or increases, enfolds or unfolds, always preserving in itself some degree of life [*vitalitas*] or, if you prefer, some degree of primitive activity [*actuositas*]. (*On Body and Force, Against the Cartesians,* May 1702, AG 253)[17]

[16] G II 111–129.

[17] It is very interesting to observe that (in addition to the connection between infinity and life) the notion of a degree of life is clearly related here to degree of primitive activity.

Leibniz's last point that some degree of life is always preserved in a natural machine, and that it corresponds to some degree of primitive activity will be developed in the last section. For now I would like to observe that, according to Leibniz, a living being is infinite both in being ever active and in its internal structure.[18] It is significant, too, that the infinity of living beings is intrinsically related to their being creations of an infinite creator, or, in other words, to their being what Leibniz calls "divine machines."

This point is made explicitly in the *Monadology* §64:

> Thus each organized body of a living being is a kind of divine machine or natural automaton, which infinitely surpasses all artificial automata. For a machine constructed by man's art is not a machine in each of its parts. For example, the tooth of a brass wheel has parts or fragments which, for us, are no longer artificial things, and no longer have any marks to indicate the machine for whose use the wheel was intended. But natural machines, that is, living bodies, are still machines in their least parts, to infinity. That is the difference between nature and art, that is, between the divine art and our art. (AG 221)[19]

To wrap up this quick survey of texts, let us note that, in the *Principles of Nature and Grace, Based on Reason* (1714) §4, Leibniz writes: "Each monad, together with a particular body, makes up a living substance. Thus, there is not only life everywhere, joined to limbs or organs, but there are also infinite degrees of life in the monads, some dominating more or less over others" (AG 208).[20]

[18] In a letter to Lady Masham from 1704 Leibniz writes: "I define an organism or a natural machine, as a machine each of whose parts is a machine, and consequently the subtlety of its artifice extends to infinity, nothing being so small as to be neglected, whereas the parts of our artificial machines are not machines. This is the essential difference between nature and art, which our moderns have not considered sufficiently" (GP III 356).

[19] «[U]ne Machine faite par l'art de l'homme n'est pas Machine dans chacune de ses parties. Par exemple: la dent d'une roue de laiton a des parties ou fragments qui ne nous sont plus quelque chose d'artificiel, et n'ont plus rien qui marque de la Machine par rapport à l'usage où la roue était destinée. Mais les Machines de la Nature, c'est-à-dire les corps vivants, sont encore des Machines dans leurs moindres parties, jusqu'à l'infini. C'est ce qui fait la différence entre la Nature et l'Art, c'est-à-dire entre l'art Divin et le nôtre.» See also Leibniz's Fifth Letter to Clarke (arts. 115, 116, in AG 344–345).

[20] See also: "But in this way I restrict composite or corporeal substance to living things alone, that is to organic machines of nature. The rest are for me mere aggregates of substances, which I call substantiated things; but an aggregate constitutes only an accidental unity" (Letter to Des Bosses, May 29, 1716, in G. W. Leibniz, *The Leibniz-Des Bosses Correspondence*, eds. and trans. B. Look and D. Rutherford (New Haven, CT: Yale University Press, 2007), 377.

3. Empirical, Mathematical, and Metaphysical Sources for Leibniz's View

This brief survey, drawn from Leibniz's early writings to his very late ones, shows that infinity figures (though in importantly different ways) in Leibniz's description of living beings throughout his career. In fact, it shows more than that. These texts also demonstrate that, at least since the *New System of Nature* (1695), Leibniz is employing infinity not only to *describe* living beings but also as one of their *defining* features, that is, as a mark of living beings.

One must wonder, though, what were Leibniz's reasons for using infinity as the distinctive mark of living beings? It turns out that Leibniz's reasons for connecting infinity and life in this strong way reveal something not only about Leibniz but also about the philosophical and scientific background in the early modern era.

As François Duchesneau has argued, in the background of Leibniz's view, there is a new approach to the study of living things, viz., that of *iatro*-mechanism. This tradition draws on two main sources: a commitment to a mechanistic description of natural things advocated by Descartes and the new discoveries made through the invention of the microscope by Malpighi in Italy, Leeuwenhoek in Holland, and Hook in England. These observations have revealed a world of minute animals (*animalcula*) within animals, which were previously invisible.[21]

In his recent book *Divine Machines*, Justin Smith argues that Leibniz's theory of organic body indeed develops out of the *anatomia subtilis* of his predecessors, but that Leibniz renders it distinctly Leibnizian, by the introduction of infinity as a way of accounting for the fundamental difference between natural machines and artificial ones. Smith writes:

> ...however much the microstructural strain of iatro-mechanism provides a starting point for Leibniz's mature conception of organic body, the German philosopher's conception nonetheless amounts to a radical departure from the earlier subtle-anatomical conception of bodies as consisting in numerous little machines. The crucial difference lies in Leibniz's introduction of *infinity* into his account of the assemblage of machines that make up the body: for him, an organic body is contrasted with a mere machine to the extent that there is literally no lower limit to its mechanical composition. Leibniz's theory of organic body indeed develops out of the *anatomia subtilis* of his predecessors, but Leibniz takes his predecessors' work and renders it, so to speak, distinctly Leibnizian, by means of the introduction of infinity as a way of

[21] See Catherine Wilson, *The Invisible World* (Princeton, NJ: Princeton University Press, 1995).

> accounting for what he takes to be not just a difference of degree of complexity between natural and artificial machines, but rather a fundamental difference in kind. (155)

While I certainly agree with Smith on the "introduction of infinity" as the distinctive feature of Leibniz's view, I think that he slightly overstates the role microscopical and anatomical research plays in Leibniz's reasoning.[22] It should be noted that there is a serious discrepancy between the microscopists' successful project of discovering smaller and smaller *animalcula* and Leibniz's claim that one of the defining features of living beings is that they have a nested a structure that develops ad infinitum. Such a claim simply cannot be supported by observation, no matter how subtle a microscope one would use. One might suppose, as some *iatro-mechanists* have, that, as our means of observation improve, smaller and smaller animals shall be further discovered. By extrapolation, one might argue that this could go on indefinitely. But the claim that the structure of a living creature is infinite cannot be based on observation, for observation itself is necessarily finite.[23] Probably no one in the early modern era knows this better than Leibniz.[24]

[22] At the end of chapter 4 of his book, Smith raises very interesting questions: "Why the infinite nestedness? We may agree with Roger that Leibniz's theory of organic body is motivated by a desire to take account of the real. But we still must ask, why does 'the real' come out looking precisely like that? The answer seems to have much to do with the state of microscopical and anatomical research in the scientific context out of which Leibniz's philosophy emerged" (235).

[23] For a similar critique, see Jeffrey McDonough's review of Justin Smith's book in the *Notre Dame Philosophical Review* (NDPR), April 14, 2012 (http://ndpr.nd.edu/news/30317-divine-machines-leibniz-and-the-sciences-of-life-2/,2012).

[24] There is some textual evidence that Leibniz himself was well aware of the fallacy of arriving at the infinite on the basis of experience. For example, in a letter to Rémond (November 4, 1715), Leibniz writes: "The author [the reference is to Du Tertter's *Réfutation* of Malbranche's system published in Paris in 1715] adds (vol. 1, 307) that in the so-called knowledge of infinity, the mind only sees that lengths can be placed end to end and be repeated as much as one would like. Very well, but this author might consider that knowing that this repetition can always be made already amounts to knowing infinity" (G III, 658–689; English translation is from http://www.leibniz-translations.com/remond1715.htm). Leibniz's point here is that the idea that the infinite is based on thinking that one could go on indefinitely, presupposes (rather than yields) the notion of infinity. While this is written in a different context, it is clear that the same argument would apply to observation of smaller and smaller animals, and it supports the point that Leibniz should not base his claim that the structure of living beings is infinite on the microscopic observations alone. More direct evidence can be found in this passage:

"Although the conservation of the animal is favored by the microscopes, nonetheless we were aware of small bodies before their discovery, and thus we were already very well able to foresee the small animals, as Democritus foresaw the imperceptible stars in the Milky Way before the discovery of the telescope." *Gottfried Wilhelm Leibniz: Philosophische Schriften*, ed. and trans. Hans Heinz Holz, 5 vols. (Darmstadt, DE: Wissenschaftliche Buchgesellschaft, 1985), V/2, 302, translation in Smith (2011), 222.

Thus it seems to me that, however exciting, the discoveries made through the invention of the microscope are not sufficient for connecting infinity and life in the strong way that Leibniz does, that is, as a mark (or a defining feature) of living beings.

There is another (and even more important) point regarding the role infinity plays in Leibniz's view that could not have been motivated by empirical considerations. The introduction of infinity to Leibniz's account of natural machines (and thus of living beings) does not regard the mere infinity of their organs. Rather, Leibniz's use of infinity in this context reveals something deeper about the nature of living beings. For Leibniz, natural machines are not aggregates or composites of infinitely many parts. Rather, natural machines are active organic units that have an infinite *structure*. Such a structure implies that the activity of living beings is informed and regulated by a law that makes their infinite structure intelligible. I will get back to this point in my conclusion, but here I would like to point out that Leibniz's use of infinity in characterizing living beings cannot be fully comprehended in light of the observations made with the newly invented microscope.

It is also important to stress that the theories of the microscope observers (especially those Leibniz refers to) are grounded in theological commitments. This concerns in particular the issue of preformation, that is, the supposition that animals' inner structure is preformed in minute form in their creation, which would account for the apparent generation and could be seen as development of their preformed nature.[25] Thus it is very telling that, in the *New System of Nature*, Leibniz notes that the "*transformations* of Swammerdam, Malpighi, and Leeuwenhoek, the best observers of our time, have come to my aid, and made it easier for me to admit that animals and all other organized substances have no beginning, although we think they do, and that their apparent generation is only a development, a kind of augmentation" (AG 140).

As Smith observes, it is also significant that Leibniz's doctrine regarding the nested structure of living beings to infinity is much more present in his philosophical writings than in his medical ones. In light of the above-mentioned points, I suggest that, in addition to the empirical (observational) source, Leibniz's view of living beings also draws on (1) a mathematical source, related to developments in mathematics, in general, and to his development of the infinitesimal calculus, in particular; (2) a theological source, concerning the commitment to the preformation of creatures; and (3) a metaphysical source, which I pursue in the rest of this chapter.

The metaphysical source for Leibniz's use of infinity in his account of living beings is grounded in the traditional connection between infinity and being.

[25] Regarding preformation, see Karen Detlefsen's chapter in this volume.

This connection is most clearly expressed in the absolute infinity and being of God. Unlike most thinkers of the period, Leibniz, as we have seen, does ascribe infinity to created beings as one of their essential features. Unlike Pascal and Descartes, Leibniz rejects the dichotomy between an infinite Creator and finite creatures, as well as the epistemological imperative (explicit in both Pascal and Descartes) that, as finite minds, we cannot (and thus should not attempt) comprehend the infinite. In contrast, Leibniz thought that the infinite need not be dreaded; rather, it should be investigated, so that the glory of God and its expression in the created world would become more apparent and comprehensible. Thus, created substances are seen as imitations of their Creator *in this respect* (infinity) as well. This point is articulated in Leibniz's response to Pascal's note on the double infinity, where Leibniz writes:

> . . . all these wonders are surpassed by the envelopment of what is infinitely above all greatnesses in what is infinitely below all smallnesses. That is, our pre-established harmony, which has only recently appeared on the scene, and which yields even more than absolutely universal infinity, concentrated in the more than infinitely small and absolutely singular, by placing, virtually, the whole series of the universe in each real point which makes a Monad or substantial unity, of which I am one. That is, in each substance truly one, unique, primitive subject of life and action, always endowed with perception and with appetition, always containing in what it is the tendency to what it will be, to represent everything else which will be.[26]

Leibniz goes on to say that this substantial unity, which is a primitive subject of life, a living mirror, is like a "diminutive God"; like God in that it is a living, active being that will never cease to act and perceive; but unlike God in that some of its perceptions are indistinct and confused.

Leibniz's response to Pascal nicely brings out the way he conceives of the relation between infinity and life. Leibniz does not dispute Pascal's description of living beings as infinite; rather, he argues that Pascal did not go far enough in ascribing infinity to living beings. Had Pascal comprehended the true nature of the organic world, he would see that infinity cuts even deeper into the very nature of things. As Frederic de Buzon noted,

> That the parts of living beings are also living beings, and this to infinity, is exactly Leibniz's conception of natural machines, whose difference

[26] Grua, 553–555; F. de Buzon, *Les Études philosophiques*, n° 4/2010, 549–556. English translation is by Lloyd Strickland; see http://www.leibniz-translations.com/pascal.htm.

from artificial machines is only that they are 'machines to the least of their parts.' (My translation)[27]

My only slight correction to this point is that, for Leibniz, it is not the *parts* of living beings that are themselves living beings but rather that each living being consists of infinitely many other living beings, not as parts (which can presumably be taken away or put together) but as essential constituents, and that their nested structure goes to infinity. This subtle difference implies that, for Leibniz, natural machines are divinely created machines, endowed with an infinite capacity to act and perceive and thus to maintain their unity as long as they act. Since activity is their very essence, once created, they would act and live forever, unless annihilated by God. In this regard, too, creatures are clearly seen as infinite.

However, it is also clear that the kind of infinity Leibniz ascribes to created beings is not the same as the absolute infinity he ascribes to God. It is also not the (quantitative) infinity he employs in mathematics. In considering Leibniz's ascription of infinity to living beings in this metaphysical context, I will try to clarify the kind of infinity he is employing in defining living beings as natural machines, machines whose nested structure develops to infinity. My suggestion is that the infinity of creatures is best understood as a middle degree between the absolute infinity of God and the quantitative infinity applicable in mathematics. In what follows I develop this point by focusing on the relation between infinity, perfection, and being (and to some extent the activity of living beings). I will show that, according to Leibniz, these notions are interestingly correlated in that they admit of similar grades and hierarchical order.

4. Perfection, Being, Infinity, and Life

Let me begin by examining Leibniz's notion of perfection. I then consider how Leibniz's notion of perfection is related to his notion of infinity. One of Leibniz's early definitions of perfection appears in an interesting piece composed during his stay in Paris:

> Perfection is an absolute affirmative attribute; and it always contains everything of its own kind, since there is nothing which limits it.

[27] "Que les parties des êtres vivants soient aussi des êtres vivants, et ce à l'infini, est exactement la conception des machines de la nature, dont la différence avec les machines de l'art est que les premières sont machines jusques dans leurs moindres parties" (*Que lire dans les* deux infinis? *Remarques sur une lecture leibnizienne,* in *Les Études philosophiques,* n° 4/2010, 547). See also *Considérations sur les principes de vie et sur les natures plastiques,* GP VI, 543. See also my "Leibniz on Artificial and Natural Machines" in *Machines of Nature and Corporeal Substances in Leibniz,* eds. Justin E. H. Smith and Ohad Nachtomy (Dordrecht, DE: Springer, The New Synthese Historical Library, 2010).

> (March 18, 1676, *Excerpts from Notes on Science and Metaphysics*; A 6.3 391, LLC 53–55)

It is interesting to observe that this definition of perfection is reminiscent of Spinoza's definition of an attribute as being infinite in kind, and distinct from the absolute infinity God (which entails all attributes or perfections). In addition, this definition of perfection suggests that there are various kinds of perfections—one corresponding to each quality or a kind of thing, where perfection is seen as a maximum (or the highest degree) of that thing or quality, such as wisdom and power. And indeed, ten years later (1686), in the first paragraph of the *Discourse on Metaphysics*, Leibniz takes up this point and makes it explicit:

> there are several entirely different ways of being perfect, [...] God possess them all together, and [...] each of them belongs to him in the highest degree. (AG 35).

As this passage shows, according to Leibniz, perfections (seen as attributes of God) admit of both variety and degrees. In the same paragraph of the *Discourse on Metaphysics*, Leibniz articulates a very interesting test for identifying a perfection:

> We must also know what a perfection is. A fairly sure test for being a perfection is that forms or natures that are not capable of a highest degree are not perfections, as for example, the nature of number or figure. For the greatest of all numbers (or even the number of all numbers), as well as the greatest of all figures, imply a contradiction, but the greatest knowledge and omnipotence do not involve any impossibility. (AG 35)[28]

Knowledge and power are notions that can be regarded as perfections since their having a highest degree is possible (maximal knowledge or power does not involve a contradiction). By implication, a notion that can have a highest degree (a maximum) can be regarded as a perfection. The fact that Leibniz regards the notions of greatest knowledge and maximal power to be consistent, unlike the notions of the greatest number and the greatest figure, suggests that he considers their maximum not as a quantity, that is, in a non-quantitative sense. If wisdom

[28] "For I believe it to be the nature of certain notions that they are incomparable of perfection and completion, and also of having the greatest of their kind (*suo genere summi*)" (A 6.3 552; LLC 179).

and power would be quantifiable as number and shape, they would involve similar contradictions. This suggests that the sense in which Leibniz is using perfection and infinity to qualify maximal knowledge and power is not quantitative. This point can be clearly seen in considering God's perfection (as well as that of his attributes, which Leibniz also calls perfections). God is the most perfect being, but he cannot be most perfect in a quantitative sense for otherwise its notion would be contradictory.[29] In this context, it seems rather clear that perfection is used in an absolute and non-quantitative sense. In this absolute sense, perfection signifies completeness that cannot be measured by (or compared with) size, length, speed, or any other magnitude.

But this is not the only way in which Leibniz is using the notion of perfection. In some well-known texts, such as "On the Radical Origination of Things," Leibniz defines perfection as a *degree of essence.* Here the notion of degree is explicit in the very definition of perfection. It is remarkable that, in this context, perfection also serves Leibniz as the principle of existence, in the sense that God would choose to realize the most perfect (possible) world (G VII 304). According to Leibniz, "The most perfect exists for perfection is nothing other than the quantity of reality."[30] For this reason, God would choose the most perfect world or the world whose degree of reality is the highest.

This notion of perfection is articulated in a number of Leibniz's texts. Here are some examples:

> Perfection is a degree of reality or essence (*perfectionem esse gradum seu quantitatem realitatis seu essentiae*). (Letter to Eckhard, A No. 148, 453)[31] Perfection is nothing but the degree or measure of positive reality (*la perfection n'étant autre chose que la grandeur de la réalité positive*). (GP VI, 613) Perfection is nothing but quantity of essence (*est enim perfectio nihil aliud quam essentiae quantitas*). (GP VII, 303)[32]

Leibniz uses this notion of perfection mainly in the context of justifying God's choice of the best among all possible worlds. According to Leibniz, the reason

[29] For details regarding this comparison, see my "Leibniz on the Greatest Number and the Greatest Being," in *The Leibniz Review* 15 (2005): 49–66.

[30] «Existe donc le plus parfait, puisque la perfection n'est autre chose que la quantité de réalité» (in J. B. Rauzy, ed. and trans.), *Recherches générales sur l'analyse des notions et des vérités, 24 thèses métaphysiques et autres textes logiques et métaphysiques* (Paris: Presses Universitaires de France, 1995), 469.

[31] See G. W. Leibniz, *Philosophical Papers and Letters,* edited and translated by L. E. Loemker, 2nd edition (Dordrecht, DE: Kluwer, 1969), 178.

[32] "The Perfection about which you ask is the degree of positive reality, or what comes to the same things, the degree of affirmative intelligibility, so that something more perfect is something in which more things worth of observation are found" (Letter to Wolff, AG 230).

to realize a possible world concerns its degree of perfection.[33] But let us take note that, if possible things have various degrees of perfection, it is clear that this degree would also be expressed in their actual counterparts in the created world. After all, this is precisely the reason they are created (or why they are chosen for actualization). Leibniz also makes it clear that "the reality of creatures is not the same absolute reality that is in God, but a limited reality, for that is the essence of a creature" (*De abstracto et concreto* [1688], A 6. 4. 990).

As we have seen, perfection serves Leibniz as a measure of reality. It is thus clear that creatures would have a lower degree of perfection than God. As Leibniz writes to de Volder,

> ...the supreme substance has poured forth his perfection as much as was permitted into the many substances that depend on him, which ought to be conceived of as individual concentrations of the universe and (some more than others) as imitations of the divinity. (Letter to de Volder, January 1705, in Paul Lodge's translation, PDF version 477)

Unlike Descartes and most other thinkers in the early modern period, Leibniz is not merely distinguishing between the absolute reality and perfection of God and the limited and imperfect nature of creatures. This distinction certainly holds for Leibniz, as creatures for him are imperfect with respect to the absolute perfection of God. But, at the same time, and as we have seen in the second section, in Leibniz's metaphysical system, creatures also admit of some degree of perfection.

When we connect this point with the distinction between various kinds and various degrees of perfections, it becomes clear that, for Leibniz, the distinction between God and creatures is not merely between a perfect Creator and imperfect creatures; rather, it is a distinction between the absolute perfection of God and a lesser degree of perfection ascribed to creatures (which, in turn, admit of various degrees of perfection as well).

Now, how does Leibniz relate this (graded) notion of perfection to his notion of infinity? In his *Meditations on First Philosophy*, Descartes assumes a very strong connection between infinity and perfection. He contrasts the infinity and perfection of God with human finitude and imperfection. Indeed, the connection between infinity and perfection seems to be so strong in this context that it almost appears as an identity (insofar as Descartes uses these terms interchangeably). Leibniz's analogy between perfection and infinity (imperfection

[33] "And this reason [for existence] can only be found in fitness, or in the degree of perfection that these worlds contain, each possible world having the right to existence in proportion to the perfection it contains" (*Monadology* §54 ; see *Theodicy* §§74, 167, 350, 201, 130, 352, 345, 354).

and finitude or limitation), however, is restricted to one sense of infinity, namely to the absolute sense of infinity, which pertains to the *Ens Perfectissimum* alone. Unlike Descartes, Leibniz distinguishes between different senses and different degrees of infinity. Whereas the infinity of God is absolute and includes all perfections, creatures are infinite in kind, that is, creatures are infinite, too, but to a lesser extent (or in a lower degree of infinity).

Quite early in his career (1676) Leibniz draws a very interesting distinction (or more precisely a threefold distinction) between three degrees of infinity. An early formulation of this distinction appears in Leibniz's responses to Spinoza—both in his summary of a conversation he had with Tschirnhaus about Spinoza's philosophy and in his annotations on Spinoza's letter (Ep. 12) to Ludwig Meyer—the so-called letter on the infinite. As we know well from the *Ethics* and from Spinoza's letter, Spinoza draws a distinction between the absolute infinity of God and the infinity in kind (suis generis) of its attributes.

In his summary of Tschirnhaus's account of Spinoza's philosophy, Leibniz writes:

> He [Spinoza] thinks that there are infinitely many other affirmative attributes other than thought and extension, but that there is thought in all of them, as there is here in extension; but that we cannot conceive what they are like, each one being infinite in its own kind, as, here, is space.

Immediately following this remark, Leibniz makes this note:

> I usually say that there are three degrees of infinity. The lowest is, for the sake of example, like that of the asymptote of the hyperbola; and this I usually call the mere infinite (*tantum infinitum*). It is greater than any assignable, as can also be said of the other degrees. The second is that which is greatest in its own kind (*maximum in suo scilicet genere*), as for example the greatest of all extended things is the whole of space, the greatest of all successives is eternity. The third degree of infinity, and this is the highest degree, is *everything* (*omnia*), and this kind of infinite is in God, since he is all one; for in him are contained the requisites of existing of all others." (February 1676, A 6.3 386; LLC 43)[34]

[34] «Ego Soleo dicere: tres infiniti gradus, infimum v.g. ut exempli causa asymptoti hyperbolae; et hoc ego soleo tantum vocare infinitum. Id est majus quolibet assignabili; quod et de caeteris omnibus dici potest; alterum est maximum in suo scilicet genere, ut maximum omnium extensorum est tosum spatium, maximum omnium successivorum est aeternitas. Tertius infiniti, isque summus gradus est ipsum, *omnia*, quale infinitum est in Deo, is enim est unus omnia; in eo enim caeterorum omnium ad existendum requisita continentur. Haec obiter annoto» (A 385; LLC 42).

In his annotations to Spinoza's letter, Leibniz makes these distinctions even more explicit:

> I set in order of degree: *Omnia; Maximum; Infinitum.* Whatever contains *everything* is maximum in entity; just as a space unbounded in every direction is maximum in extension. Likewise, that which contains everything is most infinite, as I am accustomed to call it, or the absolutely infinite. The *maximum* is *everything* of its kind (*omnia suis generis*), i.e., that to which nothing can be added, for instance, a line unbounded on both sides, which is also obviously infinite for it contains every length. Finally those things are *infinite in the lowest degree* whose magnitude is greater than we can expound by an assignable ratio to sensible things, even though there exists something greater than these things. (A 6.3 282, LLC 115)[35]

If the analogy between perfection and infinity may be developed along these lines, then I would suggest that creatures are also partly perfect, that is, perfect in a particular way—the very way that defines their unique essence and nature (i.e., their unique set of predicates or their complete concept). Unlike God, creatures have some perfections but also many imperfections and limitations.[36]

This complex and interesting status of creatures, set between the absolute infinity of God and the mere infinity of mathematical beings, requires much more attention than I have space for here. My discussion is therefore limited to drawing attention to this complex and interesting status that constitutes,

[35] «Ego semper distinxi Immensum ab Interminato, seu terminum non habente, Et id cui nihil addi potest, ab eo quod numerum assignabilem superat. Breviter gradus constituo: Omnia, Maximum, Infinitum. Quicquid *omnia* continent, est maximum in entitate; quemadmodum spatium in omnes dimensiones interminatum est maximum in extensione. Item quod omnia continet, infinitissimum, ut vocare soleo, sive absolute infinitum est. *Maximumm* est *omnia* sui generis, seu cui nil apponi potest, ut linea recta utrinque interminata, quam et infinitam esse patet; nam omnem continet longitudinem. *Infinita* denique *infini gradus* sunt quorum magnitude major est, quam ut a nobis ratione assignabili ad sensibilia, possit explicari. Quanquam aliquid detur ipsis majus. Quemadmodum spatium infinitum inter asymptoton et Hyperbolam Apollonii comprehensum quod ex moderatissimis infinitis unum est, cui respondet quodammodo in numeris summa huius spatii; 1/1 ½ 1/3 ¼ etc. quae est 1/0. Modo scilicet intelligamus ipsum 0, seu Nullitatem, aut potius hoc loco quantitatem infinite seu inassignabiliter parvam, majorem esse minoremve, prout ultimum seriei huius fractionum, infinitae, Nominatorem, qui infinitus et ipse est minorem majoremve assumserimus. Maximum enim in numeros non cadit.»

[36] "The imperfections, on the other hand, and the defects in operations spring from the original limitation that the creature could not but receive with the first beginning of its being, through the ideal reasons which restrict it. For God could not give the creature all without making of it a God; *therefore there must needs be different degrees in the perfection of things and limitations also of every kind*" (*Theodicy* §31, my italics).

I suggest, the mark of living beings for Leibniz, namely that each (created) living being is both perfect and infinite in a certain way and that this implies that it is also limited in many other ways (for it realizes one way of being, unlike God who is all things). It is like a diminutive God in being infinite and perfect in its own kind—that is, in one limited and determined way, which also defines it as a unique expression of God.

5. Conclusion

I have argued that Leibniz connects the notion of infinity and life so that, after the *New System of Nature*, infinity figures as a mark of living beings. I examined Leibniz's sources for this view, arguing that it is best seen as a synthesis of mathematical, empirical, theological, and metaphysical sources. I then focused on the metaphysical source, examining the connection between infinity, perfection, and being in some more detail. My conclusion is that Leibniz defines living beings as infinite in a qualified (graded) sense, so that they are infinite in kind, that is, infinite in some ways but limited in others, and that their infinity is distinct from the absolute infinity (and perfection) of God, on the one hand, and the mere infinity of mathematical things, on the other.

Let me now return very briefly to my original questions. Why does Leibniz introduce infinity into the definition of living beings? And what does infinity contribute to his notion of a living being? In thinking about these questions, it is important to note that, for Leibniz, a living being is divinely created, that is, a divine or a natural machine, distinct from an artificial machine, and, at the same time, an indestructible (but ever-changing, self-regulating) being. Such a (created) being is not composed of parts but has an intrinsic unity, which is derived from its inner activity and regulated by its law of generation. Such an infinite law constitutes the unique nature or essence of a living being. The activation and realization of such a law can only be initiated by an infinite creator with the result that each creature is seen as a limited (and particular) expression of God's infinite nature. Creatures, for Leibniz, are seen as diminutive gods to the extent that their nature manifests to a certain degree the absolutely infinite and active nature of the Creator.

2

What Is Life?

A Comparative Study of Ralph Cudworth and Nehemiah Grew

RAPHAËLE ANDRAULT

1. Introduction

At the end of the 17th and the beginning of the 18th centuries, vital phenomena were the focus of a violent polemic against Cartesian mechanism. Philosophers and naturalists who were engaged in this polemic proposed different conceptions of what specifies a living being. Whereas "life" was reduced to the effects of matter and motion by mechanist philosophers, it was assimilated to a special entity or power by their detractors, who wanted to defend the specificity of living phenomena vis-à-vis inanimate things and artificial machines. Thus the idea of life played an important role in the refutation of both Cartesian dualism and materialism, supposedly encouraged by mechanistic theories. This historical framework is well known and has been comprehensively studied. However, the meaning of the notion of life used in this context has not been duly examined. In this chapter, I look into this question by comparing the physico-theologies of Ralph Cudworth and Nehemiah Grew.

Indeed, in 1703, the publication by Jean Le Clerc of several chapters of Ralph Cudworth's *True Intellectual System of the Universe* (1678) and Nehemiah Grew's *Cosmologia Sacra* (1701) provoked a virulent reaction among the Republic of Letters, involving notably Pierre Bayle and G. W. Leibniz. For Bayle, Cudworth's notion of a "plastic nature" and Grew's idea of a "vital principle" were similar and should be regarded as modern versions of the Aristotelian concept of "substantial forms." Following Bayle's point of departure, this chapter examines the nature of "life" and "vital phenomena" used in Cudworth and Grew against the materialistic tendency of mechanist theories. In the first place, I see how the idea of life is used as a refutation of Cartesian dualism, that is, as a refutation of the division between *res extensa* and *res cogitans*. Then I focus on the difference between Cudworth's and Grew's conceptions of "life" and "vegetative souls" of the scholastics. Last, I show how the double apologetic ambition (against dualism and against hylozoism) is an obstacle for drawing a clear line between living and nonliving beings. My aim is to challenge the usual assimilation between what the authors of that time considered as "a body endowed with Life" and what is called today a living body.

2. Against Cartesian Dualism: Life as an Unconscious Autonomy

Cudworth's *True Intellectual System* and Grew's *Cosmologia Sacra* present two obvious common points: the apologetic ambition, which consists in demonstrating God's excellency, and the polemical background, that is, the refutation of atheism understood as the natural tendency of modern philosophical systems since Descartes, such as Hobbes's and Spinoza's. But in this double common approach, the two works proceed very differently. The *True Intellectual System of the Universe*, which is best known, is a learned but selective history of philosophy. Cudworth examines different philosophical systems, such as atomism and stoicism, in order to build a new "intellectual system" true to Christianity. It is in this context that a long digression about the hylozoism takes place at the end of the third chapter of the first book. It aims to show that what is reprehensible in hylozoism is not the assumption of a plastic nature, or a life distinct from the animal, but rather the way this plastic nature is conceived; we can admit it without danger for the religion if, and only if, it is regarded as incorporeal and, as such, an agent of God in the world. Cudworth demonstrates that a plastic nature, or a "plastic life of nature," or again a "vital energy", is an adequate tool both to account for corporeal phenomena and to apprehend rightly the way God is the cause of those phenomena.[1] As for Grew, his demonstration of the "truth and excellency of the Bible" is based

[1] There are obvious analogies between Henry More's *Immortality of the Soul* (1659; 1662), where More presents the idea of a "spirit of nature" understood as a sphere of life and activity, and Cudworth's *True Intellectual System*, where the idea of a plastic nature of life is developed. But here, we are specifically interested in the reasons for the reception of Cudworth at the beginning of the 18th century, that is to say, long after the first publication of the *True Intellectual System*. This reception by Bayle and Leibniz is focused on the question of vital phenomena, though it was just one part among others in Cudworth's monumental book. See Alan Gabbey, "Henry More and the Limits of Mechanism," in S. Hutton (ed.), *Henry More (1614–1687), Tercentenary Studies* (Dordrecht, DE: Kluwer, 1990), 19–35; and François Duchesneau, "L'idée néo-platonicienne du vivant chez More et Cudworth," in *Les modèles du vivant, de Descartes à Leibniz* (Paris: Vrin, 1998), 149–181.

on detailed descriptions of corporeal phenomena, according to a progression from simple to complex, and from inanimate to animate. The second book is about the "vital world" and examines successively the "Life," "Sense," "Mind (Phancy)," and "Intellectual Mind." Nehemiah Grew (1641–1712) was a doctor by profession, receiving his degree from the University of Leiden in 1671. He is principally known for his work on the anatomy of plants, which he presented before the Royal Society. He became thereafter the secretary of the Society—from 1677 on. In his *Anatomy of Plants* (1682), Grew explained the vegetation of plants in a mechanical way, by microstructures and motions, without any reference to an incorporeal principle. Twenty years later, in his *Cosmologia Sacra or a Discourse of the Universe. As it is the Creature and Kingdom of God. Chiefly written to demonstrate the Truth and Excellency of the Bible*, the arguments are completely different. In this book, Grew used his knowledge of anatomy to demonstrate the existence of "incorporeal principles,"[2] required to account for vital phenomena such as vegetation, sensation, and intellection. In other words, Cudworth's book is a vast philological undertaking, whereas Grew's book is first based on physiological knowledge and "goes into the detail of phenomena," according to Leibniz's statement.[3] In light of these differences, how can we explain the fact that Bayle, and then Le Clerc and Leibniz, have regarded the two books as one indistinct whole? Was Bayle fair with the two authors when he dealt with the "plastic nature" of Cudworth and the "principle of Life" of Grew as two names for the same thing and two different expressions of the same kind of thesis?

In the first place, it is true that there is a common gesture in both authors' refutation of Cartesian dualism, which consists in demonstrating the existence of an intermediate nature, between the thinking soul and the corporeal matter: a principle of life devoid of consciousness, or, according to Cudworth's innovative expression, devoid of "synesthesis," but endowed with a real power of action—a regular, efficient, and relevant action, which implies that the principle acts partly autonomously toward an end, without being forced by the blind and fortuitous interactions of the particles of matter. To show that there are a lot of incorporeal principles of life, and not only human souls, is a way to show that

[2] For a more detailed presentation, see Raphaële Andrault, "Entre anatomie et théologie: l'organisme chez G. W. Leibniz et Nehemiah Grew," in *Natur und Subjekt. IX. Internationaler Leibniz-Kongress*, eds. Herbert Breger, Jürgen Herbst, Sven Erdner (Hannover, DE: Hartmann, 2011), I, 18–26.

[3] To Burnett, 1703: "J'ai le *Wisdom of God* de Mr Ray, et la *Cosmotheorie* de Mr. Grew. Ce sont de bons livres, car ils entrent dans le détail de la nature et ne se contentent point des raisons générales, dont je ne fais pas trop de cas, quand elles ne sont point écrites d'une manière démonstrative" (*Die philosophischen Schriften von G.W. Leibniz*, ed. C. I. Gerhardt [Hildesheim, DE, and New-York: Georg Olms Verlag, 1978], III: GP III, 291).

God is everywhere in the world. This polemical upshot is particularly clear in Cudworth:

> The Occasion of which Miscarriage [of Descartes] hath been already intimated, namely from the acknowledging only *Two Heads* of Being, *Extended* and *Cogitative,* and making *the Essence of Cogitation* to consist in *Express Consciousness*; from whence it follows, that there could be no Plastick Nature, and therefore either all things must be done by *Fortuitous Mechanism,* or else God himself be brought Immediately upon the stage, for the solving of all *Phaenomena.* Which Latter Absurdity, our Philosopher being over careful to avoid, cast himself upon the Former, the banishing of all *Final* and *Mental* Causality quite out of the World.[4]

With this purpose, the refutation of Cartesian's dualism, Cudworth and Grew's demonstrations use the particular phenomena in nature that "transcend the powers of mechanism,"[5] as motion of respiration in the animals, to show that there is an intermediate principle first between divine action and natural things, and second between human mind and inert matter. According to Cudworth, plastic nature, which forms plants and animals, is said to be devoid of animal imagination and self-perception:

> There is... another *Imperfection* to be observed in the *Plastick Nature,* that as it doth not comprehend the Reason of its own Action, so neither is it *Clearly* and *Expressly Conscious* of what it doth.[6]

As examples, Cudworth invokes respiration and the pulsations of the heart. The aim of Cudworth's digression about plastic nature is indeed to show that there is "some Lower Life from the Animal"[7]—an animal life, which is apparently by

[4] *The True Intellectual System of the Universe: the first part, wherein all the reason and philosophy of atheism is confuted and its impossibility demonstrated* (London: Printed for Richard Royston, 1678), book I, ch. III: TIS, I, III, 38, 175. About Cudworth's conception of the action of God, see Justin Smith, "The Leibnizian Organism between Cudworth's Plastic Natures and Locke's Thinking Matter," in François Duchesneau and Jérémie Griard (eds.), *Leibniz selon les Nouveaux Essais sur l'entendement humain* (Montreal: Bellarmin-Vrin, 2006), 129–140.

[5] See Cudworth, TIS, I, III, 37, §3, 148: "Now to assert... that all the effects of nature come to pass by *Materiel* and *Mechanical Necessity,* or the mere *Fortuitous Motion* of *Matter,* without any Guidance or Direction, is a thing no less Irrational than it is Impious and Atheistical. Not only because it is utterly Unconceivable and Impossible, that such Infinite Regularity and Artificialness, as is every where throughout the whole World, should constantly result out of the *Fortuitous Motion* of *Matter;* but also because there are many such particular *Phaenomena* in nature, as do plainly transcend the *Powers of Mechanism,* of which therefore no sufficient mechanical reasons can be devised, as the *Motion of Respiration* in Animals."

[6] TIS, I, III, 37, §15, 158–159.

[7] TIS, I, III, 37, §26, 172.

definition endowed with consciousness. In Grew, we find also the proofs of the unconscious nature of life:

> [I]t is the same Vegetable Life, whereby these Operations are performed in Animals, as well as in Plants. And which are still going on, while we are asleep, and all our Senses locked up. And is also the reason, that while we are awake, we feel none of all those Motions, which are continually made, in the disposal of the Corporeal principles subservient herein.... [The motion of the Guts is a] Motion, whereof we have no perception, as depending on the Vegetable Life in an Animal. As also doth, the Peristaltick Motion of the Arteries.[8]

Thus, "life must occupy an intermediate position between mechanism and mind—an operative principle, without, nevertheless, any perceptive role,"[9] as emphasized by G. Giglioni. And indeed, it is within the polemical context of the refutation of Cartesian dualism in the late 17th century that life and vitality have become scientific and philosophical objects in themselves. In contrast, "life" is nearly absent from medical treatises of that time. In medical books such as Nicolas Tulp's *Observationes medicae* (1658) or Kerckring's *Spicilegium anatomicum* (1670),[10] the notion of "life" is only used when diseases call into question the continuity of the existence of a patient in case of great pain or anorexia. In other words, life, which means the continuity of an individual existence indicated by the retention of cardiac motions and respiration, usually appears only when the line between the two states "life" and "death" does not seem obvious anymore for the physician. For the anatomists, life itself is rarely conceived as an object of scientific examination or explanation, but only as a vulgar notion reducible to some more fundamental causes. Life is the *explanandum*, not the *explanans*: an effect, and not a cause or a principle. For instance, in Steno's *Elementorum Myologiae Specimen* (1667), the signs of "life" and "death" are only exterior manifestations of inner motions like those of the motor fiber. The proper matter of the anatomist in this case is the motion of the fibers, that is to say, the cause of the motions of the heart and respiration, but not the signs of life and death, whose boundaries are determined by common sense. The adjectives

[8] Nehemiah Grew, *Cosmologia Sacra or a Discourse of the Universe. As it is the Creature and Kingdom of God. Chiefly written to demonstrate the Truth and Excellency of the Bible, in five books* (London: Printed for W. Rogers, S. Smith, and B. Walford, 1701), book two, ch. 1: CS, II, I, 36.

[9] G. Giglioni, "Automata compared. Boyle, Leibniz, and the debate on the notion of life and mind," *British Journal for the History of Philosophy*, vol. 3 n° 2 (1995): 249–278, 274.

[10] *Observationes medicae* (Amstelodami: apud Ludovicum Elzevirium, 1652), liber 1, cap. XIX, 42; and *Spicilegium Anatomicum* (Amstelodami, 1679), Obs. 80, 156: "inde exsugebat per anni totius spatium, sustentavit vitam, si languor ille, & illam paulatim, consumens macies, qua tandem extincta est, non potius mors diuturna, quam vita dicenda est."

"vital" or "active" are completely absent from this book.[11] Beyond Steno, at that time, life is usually defined by the anatomists, according to Descartes's view, as the effect of the circulation of the blood,[12] of the agitation of spirituous liquors,[13] or of the perpetual motion of particles.[14]

3. Against Materialism: Life versus Bodies

If life is a third causal agent in the world, between human soul and inert matter, we could believe, as Bayle does, that Cudworth's "plastic nature" or Grew's "vital principle" are only some modern versions of Aristotelian "substantial forms":

> Mr. Cudworth and Mr. Grew, very great philosophers..., did not find that it was worthy of them to strengthen and to clarify the Cartesian hypothesis, which is in the end the most capable of supporting the spirituality of God: they found more glory in strengthening the uncertain and almost appalling sect of Peripaticians, I mean in updating and repackaging the doctrine of substantial forms, the one by putting forth the system of the plastic faculty, the other by supposing a vital world that is distinct from the material world.[15]

[11] Contrary to Mirko Grmek's statement, for whom Steno's motor fiber is an "active fiber" in *La première révolution biologique: Réflexions sur la physiologie et la médecine du XVIIe siècle* (Paris: Payot, 1990), 172. See Steno, "Elementorum Mylogiae Specimen," in *Opera philosophica*, ed. Vihelm Maar (Copenhagen: V. Tryde, 1910), II, 105: "Adde, quod de fibra motrice agatur, de parte, quae membra agitat, quae aerem inspirit, quae sanguinem movet, paucis, unde vitae mortisque signa dependent."

[12] See "L'Homme," in *Oeuvres de Descartes*, eds. C. Adam, P. Tannery, B. Rochot, and P. Costabel (Paris: Vrin-CNRS, 1964–74), XI, 202: "[il ne faut concevoir] en cette machine [l'homme] aucune autre âme végétative, ni sensitive, ni aucun autre principe de mouvement et de vie, que son sang et ses esprits, agités par la chaleur du feu qui brûle continuellement dans son cœur, et qui n'est point d'autre nature que tous les feux qui sont dans les corps inanimés."

[13] William Croone, "On the reason of the movement of the muscles," trans. Paul Maquet, intro. Margaret Nayler, *Transactions of the American Philosophical Society*, vol. 90 (2000), with the facsimile of the original edition in Latin (ed. by F. Hayes, London, 1664), §12, 70 : "Spirituosis hujusce generis liquoribus, singula in Animantis corpore partes vehementer turgent.... Omnes autem ubicunque sunt, caloris nativi ac circulationis beneficio, in constanti motu ac agitatione existant: & haec ipsa Agitatio, id ipsum est quod Vitam appelamus."

[14] Marcello Malpighi, "De Gallis" (1679), in *Opera omnia* (London: Robert Scott, 1686), II, 17: "Animalium vita, cum perpetuo effluvio manutenatur; talem corporis compagem exigit, qua jam recepta largo dispendio reddet; alienaque denuo, & quotidie admittat: ita ut vitae chorea continuato introitu & egressu perennet."

[15] "Continuation des Pensées diverses écrites à un Docteur de Sorbonne, à l'occasion de la Comète qui parut au mois de Décembre 1680, ou Réponse à plusieurs difficultés que Mr. *** a proposées à l'Auteur," in *Pensées diverses* (Rotterdam, 1705), 90. "M. Cudworth et M. Grew très grands Philosophes (...) n'ont pas trouvé qu'ils fût digne d'eux de fortifier et d'éclaircir l'hypothèse

According to Bayle, what is dangerous in this undertaking is to ascribe to bodies a kind of motor force and autonomy, which would imply a compatibility between the two and, as a consequence, an autonomy of matter vis-à-vis God.[16]

The comparison of "substantial forms" with vital principles could be true if vital principles were inherent to matter or, at least, animated matter from the inside. Even though Cudworth evokes an "inward principle" for apologetic reason, he has to insist on the absolute heterogeneity between life, which is something mental, and corporeal bodies. In both cases, there is no rehabilitation of the scholastic couple form/matter, considering the very sharp opposition between, on the one hand, incorporeal principles, which are necessarily exterior to matter because they have a divine origin, and, on the other hand, the properties of matter, which are dependent, passive, and, for Cudworth only, erratic and resistant to the order introduced by the laws of nature. Indeed, Grew's explicit starting point is the fact that God has life, and God is incorporeal. Thus, for Grew, life has to be incorporeal. And if there is something as Life in nature, it is a proof of the divine creation of nature—living things are the marks of the creator in the creation. In the introductory lines of *Cosmologia sacra*'s second book, which is devoted to "Life," we read:

> The Universe consisteth of the Corporeal and the Vital World: the Latter of which, is next to be considered. And first, I shall prove, That there is a Vital Substance in Nature, distinct from a Body.... God, who as he is the Cause of all other Beings, is the most Substantial Being: is Himself a Substance Vital and Incorporeal.[17]

Cartesienne, qui est dans le fond la plus capable de soutenir la spiritualité de Dieu: ils ont trouvé plus de gloire à fortifier la secte chancellante et presque atterrée des Péripatéticiens, je veux dire à mettre dans un plus beau jour et sous une nouvelle face, la doctrine des formes substantielles, l'un en illustrant le système de la faculté plastique, l'autre en supposant un monde vital distinct du monde matériel."

[16] *Ibid.* "La forme plastique de M. Cudworth, et le principe vital de M. Grew sont cependant dans le même cas, et ainsi ils ôtent à cette objection contre les athées toute sa force. Car si Dieu a pu donner une semblable vertu plastique, c'est une marque qu'il ne répugne point à la nature des choses qu'il y ait de tels agents, ils peuvent donc exister d'eux-mêmes, conclura-t-on. Vous comprendrez ceci par une comparaison. Si la matière peut recevoir de Dieu la force motrice, il y a une compatibilité naturelle entre la matière et la force motrice. On peut donc supposer également et que la matière existe par elle-même, et que la vertu motrice lui est propre essentiellement. Ceux qui supposent comme la plupart des Cartesiens que la matière est incapable d'être investie de la force de se mouvoir, et que Dieu seul peut produire le mouvement sont beaucoup plus en état de démonter les athées."

[17] CS, II, I, 31.

As regards Cudworth, we find the same kind of assimilation between life and incorporeity:

> But though the Plastick Nature be the Lowest of all Lives, nevertheless, since it is a *Life*, it must needs be *Incorporeal*; all Life being such.[18]

Certainly, Grew's refutation of materialism is different from Cudworth's. Most notably, Grew admits a Cartesian conception of body[19] : Bodies are defined by mechanical properties as figures, structures, and sizes. So, by hypothesis, if corporeal modifications could produce vital phenomena, there would be three options, all of which Grew refutes: (1) It could be by the "subtlety" of the particles. But size cannot change the nature of bodies: "Body, being subtilized, can loose nothing of its Corporeity."[20] (2) It could be by their organization, either by their figures (but a square is not more vital than a triangle because every square contains two triangles at least); by the combination of the parts (but a watch is not more "vital" when its parts do not "lie in a confused heap"); or by their motions (but a paper kite is not more vital when it is moved by the wind). (3) Life could be caused by a special property of the body, but this supposition is perceived by Grew as absurd. According to this demonstration, no mechanism accounts for the manifest difference between animate and inanimate. So, we have to admit an incorporeal principle of Life.

In Cudworth, the refutation of materialism proceeds differently. Cudworth's matter, contrary to Grew's matter, is not really or faithfully Cartesian. First, matter is for him "dead" and passive. Second, matter is submitted only to fortuitous motions. As a consequence, matter is *resistant* to any regularity or any ordering by the laws of nature. To account for the regularity of phenomena, we have to then conceive an "immediate agent" of God, like a plastic nature, by which "the things of nature are administered":

> Wherefore, the *Divine Law* and *Command*, by which the things of Nature are administered, must be conceived to be the Real Appointment of some *Energetic, Effectual,* and *Operative Cause* for the Production of every Effect.[21]

[18] TIS, I, III, 37, §20, 163.

[19] As emphasized by Brian Garrett, "what makes Grew interesting is his simultaneous commitment to atomistic and mechanistic explanation regarding matter, along with the somewhat panpsychist view that all material bodies, but certainly vegetative bodies, are associated with a vital principle" ("Vitalism and Teleology in the Natural Philosophy of Nehemiah Grew," *British Journal for the History of Science*, 36 (2003), 63–81, 3).

[20] CS II, I, 32.

[21] TIS, I. III, 37, §2, 147.

Nevertheless, despite these differences, both authors maintain a dualistic system, opposing mental and corporeal, to avoid promoting materialism.[22] From this point of view, Le Clerc is right to criticize Bayle's statement:

> Mr. Cudworth and Mr. Grew do not say that God placed a motive force in matter, which exists there as an accident or as a property in its subject. They strongly attack this thought when they maintain that matter is incapable of life, and they say that the *plastic and vital natures* are immaterial substances. Thus what [Bayle] says here does not concern them, and I have no need to respond.[23]

To summarize, the refutation of materialism, whether it be mechanist or hylozoist materialism, implies in each case two features of "life": (1) Life is associated with a mental and final activity, and defined in contrast to "dead" matter; and (2) life does not necessarily imply a conscious and intelligent activity, in contrast with Cartesian soul. Life is defined negatively in two ways:

— In a *generic* sense, life is just opposed to body as an equivalent to "mental." It is even possible to ascribe some natural properties of the soul to life itself, as in this passage from Cudworth: "if the Souls of Men and Animals be at any time without *Consciousness* and *Self-perception*, then it must needs be granted, that Clear and Express *Consciousness* is not Essential to Life."[24] Insofar as Cudworth and Grew want to refute materialism, they have to give to life a broad meaning, a generic sense, according to which everything that is not corporeal, God included, is vital: They cannot admit that God has only a vegetative Life, devoid of sense and understanding.

— In a *specific* sense, life is just the lowest degree of what is incorporeal. In this last sense, life is equivalent to "vegetation" or to the vegetative Life, because it is devoid of sense, consciousness, and understanding. Generally speaking, life means the motions of vegetation, that is to say growth, reproduction, circulation and, in one word, all inner regular, continuous,

[22] See John Henry, "The Matter of Souls: Medical Theory and Theology in Seventeenth-Century England," in R. French and A. Wear (eds.), *The Medical Revolution of the Seventeenth Century* (Cambridge, UK: Cambridge University Press, 1989), 87–113, 88; and Yves-Charles Zarka, "L'être et l'action selon Ralph Cudworth," *Philosophie* (1997), n° 53, 57–69, 60.

[23] *Bibliothèque choisie* (Amsterdam: Schelte, 1705), V, 300–301. "Mrs Cudworth et Grew ne disent point que Dieu ait mis dans la Matière une force motrice, qui y existe comme un accident, ou comme une propriété dans son sujet. Ils attaquent fortement cette pensée, lorsqu'ils soutiennent que la Matière est incapable de Vie, et ils disent que les *Natures plastiques et vitales,* sont des substances immatérielles. Ainsi ce [que Bayle] dit ici ne les regarde pas, et je n'ai que faire d'y répondre."

[24] TIS, I, III, 37, §17, 160.

and unconscious motions that seem to make the difference between living beings and inert things. According to Grew, "the lowest Species of Life, whereof we can have any conception, is such a sort of Life, as is without Sense." [25]

It seems that when Cudworth uses the specific meaning of life, life as vegetation, he employs the words "a Plastick Nature," "the Plastick Nature," or "the Plastick Life of Nature," whereas he uses the term "life" when he means the generic sense of life, life as something mental. It is quite consistent with the following citations. First, he clarifies: "the *Vegetative Soul* [of the Platonists and Peripateticks] is no other than a *Plastick Nature*"; and speaks about the "*Vegetative* or *Plastick Power* of the Soul."[26] And then:

> We conceive, that the first Heads of Being ought rather to be expressed thus; *Resisting* or *Antitypous Extension*, and *Life* (i.e. *Internal Energy* and *Self-Activity*), and then again, that *Life* or *Internal Self-Activity* is to be subdivided into such as either acts with express Consciousness and *Synaesthesis*, or such as is without it; the Latter of which is this *Plastick Life* of Nature: So that there may be an Action distinct from Local Motion, or a Vital Energy, which is not accompanied with that Fancy, or Consciousness.[27]

Or:

> [T]his *Plastick Nature* is so far from being the *First* and *Highest Life*, that it is indeed the Last and Lowest of all *Lives*, it being really the same thing with the *Vegetative*, which is Inferiour to the *Sensitive*.[28]

Insofar as life means sometimes the lowest degree of the triad of vegetation, sense, and thought, we could find a similarity with the scholastic triad of "vegetative, sensitive and rational souls." This similarity is even emphasized by Cudworth when he compares his plastic nature with the vegetative soul. However, it is only an apparent similarity. In Aristotle, every living being is endowed with a vegetative soul. It is not the case for Cudworth: Although some nonliving beings (the universe, the terrestrial globe, a mineral) have a plastic nature, some living bodies, like plants, do not:

[25] CS, II, 1, 35.
[26] TIS, I, III, 37, §6, 153; and §12, 157.
[27] TIS, I, III, 37, §16, 159.
[28] TIS, I, III, 37, §19, 162–163.

> Though it be no reasonable to think, that every Plant, Herb, and Pile of Grass, hath a Particular Plastic Life, or Vegetative Soul of its own, distinct from the Mechanism of the Body, nor that the whole Earth is an Animal endued with a Conscious Soul; yet there may possibly be, for ought we know, one Plastick Nature or Life belonging to the whole Terrestrial globe, by which all Plants and Vegetables, continuous with it, may be differently formed, according to their different Seeds, as also Minerals and other Bodies framed, and whatsoever else is above the Power of Fortuitous Mechanism effected.[29]

For Cudworth, it is more reasonable to admit one unique plastic nature for the whole earth than to give to every living being its own plastic nature. This means that plastic nature is not necessarily an individual principle, but may be an unconscious substance that infuses into the different parts of corporeal nature. In this respect, it does not seem true that Cudworth gives to each individual a plastic nature or a substantial form conceived as an immaterial active force.[30] Leibniz, for example, was well aware of this difference between Cudworth's system and his own, which gives to each (living) individual its own principle of life or soul: In 1689, when annotating *The True System*, he remarks that contrary to Cudworth's assumption, every plant deserves its own soul.[31] According to Cudworth's quotation, the vegetation of plants (their growth for instance) may be explained in ascribing a plastic nature to the terrestrial globe, which could communicate its formative power to the plants through their roots. But in this case, how are we to explain the fact that some plants keep growing once picked and put into water? For Leibniz it is far more reasonable to ascribe to each plant its own life. Life is a property of individuals, and not a diffused power that can be communicated by contact or contiguity. Later on, in his *Considérations sur les principes de vie et les natures plastiques* (1705), Leibniz assimilates "life" to the soul, or something analogous: "Life" is not only the property that requires the individuation of things, or an individual as its bearer, but also a principle that

[29] TIS, I, III, 37, §25, 171.

[30] See Pauline Phemister: "[According to Cudworth, the] individual living being has two aspects: the immaterial active force and the matter or body upon which this active force « acts » and with which it forms a living being.... The position is also remarkably close to Leibniz's own Aristotelian understanding of corporeal substance as a combination of soul or substantial form and organic body." ("All the time and everywhere everything's the same as here: The Principle of Uniformity in the Correspondence between Leibniz and Lady Masham," in Paul Lodge (ed.), *Leibniz and his Correspondents* (Cambridge, UK: Cambridge University Press, 2004), 193–213, 195).

[31] See G. W. Leibniz, *Sämtliche Schriften und Briefe* (Berlin, Akademie-Verlag, 1923– [?]), VI, vol. 4, B_2, 1950: "Et postera dicit etiamsi cuilibet plantae non necesse sit assignare talem vitam plasticam, sufficere ut sit in globo terrae cui plantae cohaerent [+ sed plantae eam habent et avulsae et in vasis positae+]."

individualizes things, that is, which makes a thing *one* individual substance. In one word, for Cudworth, a plastic nature can be a transmissible power that can be shared, whereas for Leibniz, it should be an individual property—a principle peculiar to truly indivisible individuals.

4. The Properties of Life

As we have seen, the apologetic ambition of praising God's creation and proving the truth of Christianity is common to Cudworth and Grew. This explains why for both, "life" has two dimensions: a generic one, life as something mental, to refute the materialism, and a specific one, life as vegetation, to prove God's presence everywhere throughout nature. Nevertheless, the two authors do not ascribe to "life" the same properties and powers. To better understand some of the differences between the two authors, we have to examine what life, or plastic nature, means exactly for each of them, and of what life is capable. We cannot find any positive and exhaustive definition of life in those two books, particularly in Grew, but there are still some descriptions, equivalences, and features associated with the notion.

For Cudworth, life has essentially four features and functions: (1) A plastic nature is something that can directly give motions and directions to bodies. It is described as an internal energy, and even a self-activity. We may indeed identify life as a "motor force" in this sense. Quoting Plotin (Enneads, 3, 2, 16): "[E]very Life is Energy, even the worst of Lives, and therefore that of Nature."[32] This cause or energy produces among other things the motion of respiration in the animals. (2) As a plastic nature, the second property of life is the formation of bodies and their growth. Cudworth speaks about something superior to mechanism, which "restores Flesh that was lost, consolidates dissolved Continuities, incorporates the newly-received Nourishment, and joyns it Continuously with the preexistent parts of Flesh and Bone." But he also esteems that the same plastic nature builds the frame of "mineral and other bodies."[33] This is the precise meaning of "plastic," that is, *formative power.* (3) A general plastic nature of the whole universe is responsible for the *unity of the world,* that is to say, for the "uniform Order and Harmony."[34] (4) It is life also that explains the functioning of our soul, for instance in our *reasoning*: "the Rational Life

[32] TIS, I, III, 37, §16, 159.

[33] TIS, I, III, 37, §22, 167.

[34] TIS, I, III, 37, §20, 164: "The same is to be said likewise concerning the Plastick Nature of the whole Universe, in which... *all things are ordered together conspiringly into one.* It must be one and the same thing, which formeth the whole, or else it could never have fallen into such an Uniform Order and Harmony. Now that which is One and the Same, acting upon several distant parts of Matter, cannot be Corporeal."

that is in us ought to be accompted a much Higher and more Noble Perfection than that *Plastick Life of Nature*."[35] Or also: "There is also another more Interior kind of *Plastick Power* in the Soul (if we may so call it), whereby it is Formative of its own Cogitations, which it self is not always Conscious of."[36]

As a consequence, life, or at least the different sorts of life in nature, can be responsible for a large range of actions: Life is capable of every action that seems beyond the powers of matter and motions. And this diversity of actions implies a variety of bearers of Life. The plastic nature can be either a faculty of a soul, whether animal or human, or directly a soul itself. This is a question that is left undetermined. Cudworth writes, for instance: "[Life lower than the Animal] is either a Lower Faculty of some Conscious Soul, or else an Inferior kind of *Life* or *Soul* by itself; but essentially depending upon a Higher Intellect."[37] But, as we have seen, there should be also a general plastic nature for the whole universe in order to explain the harmony of the whole:

> Besides this Plastick Nature, which is in Animals, forming their several Bodies Artificially, as so many Microcosms, or *Little Worlds*, there must be also a general Plastick Nature in the *Macrocosm*, the whole Corporeal Universe that which makes all things thus to conspire every where and agree together into One Harmony.[38]

In this last sense, life cannot be assimilated to the soul without risking encouraging the view according to which the world is a unique living animal.

As for Grew, he defines "life" as a "sphere of activity" dependent upon a vital substance. As the same time, he maintains the Cartesian thesis according to which the quantity of motion is always the same in the universe. This means that vital principles cannot move bodies immediately, but can only give directions, and directions suitable to the corporeal organization of bodies:

> Yet neither by this Life, nor the Subject of it, do I mean, a Principle of Motion. The Universal Stock of Motion, as that of Matter, being neither encreased, nor diminished, but only transfer'd. But I mean, a certain Power to determine the manner of its being transfer'd; or of returning an Impression upon Bodies, suitable unto that which it receives.[39]

We can deduce from this that "life" is the cause of motions that are specifically "vital" only because they are not immediate, automatic, or mechanical reactions

[35] TIS, I, III, 37, §19, 162.
[36] TIS, I, III, §17, 161.
[37] TIS, I, III, 37, §26, 172.
[38] TIS, I, III, 37, 167.
[39] CS, II, I, 35.

to stimulations: Vital actions, that is actions caused by a vital principle, are reactions that do not seem to be the mere effect of a mechanical impact insofar as they are appropriate or "suitable" to the circumstances. This vital action can be the reasoning of our soul or the motions of the animals. In contrary to Cudworth, Grew stays relatively close to Descartes's legacy. For Grew, life cannot be a spring of motion. So Bayle and Leibniz were quite wrong in their critique of Grew, as he never clearly defended an interaction between incorporeal principles and bodies.

5. What Is Alive and What Is Not? Some Difficulties

Commentators often wonder about the definition of life, that is, about the *comprehension* of the concept, but they neglect the question of its *extension*. Of what kind of things can "life" be predicated? It is often assumed that those things are the very things we consider these days as "living beings." Nevertheless, there is no exact equivalence between, on the one hand, phenomena that belong to living beings, or at least to what we today call living beings, and, on the other hand, phenomena that allow Cudworth and Grew to illustrate the properties of life.

As we have seen, for Cudworth and Grew, vital phenomena are first and foremost the vegetation in plants, as well as the several sorts of animal and human unconscious activities. But, and it is something that has not been often underlined, life, vital substances, or plastic nature have also a "sphere of activity" in the so-called inanimate world. What is "animated by life" is broader, strictly speaking, than the group of all living beings. For Cudworth, there is something similar between a mineral and a plant, and this is why we need a principle like a plastic nature endowed with life to explain their formations. Cudworth needs also to invoke a plastic nature of the whole universe to explain its unity and harmony. This implies that there is a life everywhere, as the "fortuitous power of mechanism" is unable to account for the phenomena, including in the inert parts of the world, insofar as they exhibit some continuity, regularity, and complexity. The extension of the group of things that are completely inert and dead is rather fuzzy. The lack of systematicity is emphasized by the adjective "particularly" [*surtout*] in Le Clerc's following statement: "Mr. Cudworth believes that bodies, particularly those which are organized, are not formed by mere matter moved according to the laws of motion."[40] Then, for Cudworth, some bodies, which are

[40] Le Clerc, *Bibliothèque choisie*, V, 1705, 292: "Mr Cudworth croit que les Corps, *surtout* ceux qui sont organisés, ne sont pas formés par la pure Matière agitée selon les lois du mouvement, comme Descartes l'a cru."

for us not organized or living, are formed by a plastic nature or the plastic life of nature, contrary to Leibniz's thesis that only organized bodies are endowed with principles of life (regarded as perceiving and desiring souls): "principles of life belong only to organic bodies" (*"les principes de vie n'appartiennent qu'aux corps organiques"*).[41]

In Grew's *Cosmologia sacra*, life is required to explain the cohesion of bodies. Strikingly enough, the first illustrations that Grew gives of the "power of life" are colors, odors, magnetic inclination, and gravitation. Life as a "sphere of activity" is that "also, wherewith every Corporeal Principle may be probably Endow'd: or, to speak properly, may be Animated." Thus it seems that, in these two books, there is no clear-cut line between what is generally regarded as something inanimate and what is generally regarded as living. And, indeed, if Life, vital substances, or plastic nature have to testify for the presence of God throughout nature, it might be problematic to limit divine expression to plants, animals, and human beings—especially if the aim is to demonstrate the unity and harmony of the universe.

As a matter of fact, the apologetic ambition of the two authors explains (1) why the line between living and nonliving is so hard to draw, and (2) why there seems to be a discrepancy between what "life" is supposed to explain and the properties attributed to it, that is to say, between vital phenomena as an *explanandum* and "life" as an *explanans*. The apologetic ambition requires a demonstration of the immateriality of life. In this respect, two aspects of life are emphasized: the regularity of bodies and the whole nature, whose formation supposes *intellection* and *consciousness*, and the faculties of animals such as *sensation*. These two aspects, intelligence and sensation, clearly exceed the properties of matter. And yet "life" cannot be identified only with intellection and sensation. If it were the case, there would be no proof that life or vital energy is spread throughout nature whose unity is manifest. So, there would be no proof that, first, life goes past the realm of animals and human souls and, second, that God is not immediately the creator of every living being. In this last case, the danger would be to support the Cartesians instead of refuting them. Thus, mental features of life are needed to refute materialism, but non-mental features of life are the best proof of the omnipresence of vital principles that are God's agents. By "non-mental features," we mean vegetative actions, as respiration or growth, that are too poor or limited to be direct productions of God himself but too complex to be only the result of a material mechanism.

The association of those mental and non-mental features raises some difficulties. In the first place, according to the generic meaning of life, sense and thought are sometimes depicted as potentialities of life in order to refute materialism.

[41] GP VI, 539.

But according to the specific meaning, life does not necessarily imply sense and thought. This may cause a problem in the comprehension of the notion. Let's attend to what Grew writes:

> Were Life, an immediate Adjunct of Body, as Motion is; then as all sorts of Bodies are capable of all sorts of Motion; so they would be capable of all sorts of Life; even of Intellection itself. That is, every Atom, as an Atom, would be capable of being Intellectual. To avoid which Absurdity, we must allow the being of a Substantial Principle, distinct from Body, as, the proper and immediate subject of Life.[42]

It is difficult to understand why life as a property of matter could induce confusion among vegetative, sensitive, and rational faculties, whereas a substantial principle called "vital" would enable us to understand the real distinction among these three kinds of life.

In the second place, according to Grew, "the difference between Vegetable and Sensible Life, seemeth to be this; that in the former, the Impression arising from the Motion, is entirely reflected, and lost in an Instant. In the latter, it is immersed and retein'd."[43] So, vegetable life acts as a reflex. But in this case, in what sense exactly does this reflex differ from a mere mechanical reaction? The idea of an immaterial source of life, required to conceive why a living being does not react merely mechanically to the motions he receives, is clearer for the superior sorts of life (i.e., sense and thought) than for the vegetable life. In this last case, it is hard to grasp the specific skills attributed to life versus bodies.

In the third place, it is difficult to understand in what sense vital principles or the plastic nature of life may account for phenomena that are explained neither by God's immediate action, nor by the actions of a soul endowed with intellection like the human soul. The *explanans* does not seem to meet the *explanandum*. Indeed, in Cudworth, it is said that life or plastic nature can be devoid of intellection and knowledge. But at the same time, it is because matter is devoid of intellection that a plastic nature or a life is needed in nature. More precisely, a plastic nature is able to build a body, whereas matter does not: "For is it not altogether as Absurd and Ridiculous, for Men to undertake to give an accompt of the Formation and Organization of the Bodies of Animals, by mere Fortuitous Mechanism, without any *Final or Intending Causality*."[44] And yet, a plastic nature is devoid of understanding, which seems required to build a body, or even to direct its motions: It operates "Fatally and Sympathetically,

[42] CS, II, I, 34.
[43] CS, II, II, 37.
[44] TIS, I, III, 37, §3, 149.

according to *Laws* and *Commands,* prescribed to it by a Perfect Intellect, and impress upon it."[45] How would an immaterial vital principle devoid of intellection be more able to build an organized body than matter itself directed by God's laws?

This fits one of the main criticisms of Bayle and Leibniz. For Bayle: "Nothing is more awkward for the atheists than to find themselves reduced to attributing the formation of animals to a cause that does not have an idea of what it is doing, and that regularly executes a plan without knowing the laws that it is executing."[46] Generally speaking, "a being that neither conceives what it is doing, nor the plan that it is to carry out, nor the goal toward which it is to tend, could not do anything in a regular fashion"[47]: Regularity of the phenomena requires intellection and conscious finality as its cause. As for Leibniz, according to his *Éclaircissement sur les principes de vie et les natures plastiques,* Cudworth's plastic nature does not satisfy the principle of economy:

> As for these immaterial plastic substances, I do not well see how they suffice here, nor in what respect they have the advantage over mechanism. It seems that they would need a completely divine intelligence in order to form the organs of animals with the knowledge of an architect: but if they do not have this architectonic knowledge, and if it is necessary that God intervene here and direct things in particular, then what good are these intelligences?[48]

This difficulty is intrinsic to Cudworth's system. As energy, self-activity, or sphere of activity, vital substances or plastic nature seem to be autonomous. It is what distinguishes bodies from life or from things endowed with life. But at the same time, in order not to support atheism, this autonomy has to be conceived as something still dependant upon a divine activity. This is partly why it is so

[45] TIS, I, III, 37, §26, 172.

[46] *Continuation,* 90–91. "Rien n'est plus embarrassant pour les athées que de se trouver réduits à donner la formation des animaux à une cause qui n'ait point l'idée de ce qu'elle fait, et qui exécute régulièrement un plan sans savoir les lois qu'elle exécute."

[47] *Histoire des ouvrages des Savans,* August 1704, 384. "Un être qui ne conçoit point ce qu'il fait, ni le plan qu'il doit remplir, ni le but où il doit tendre, ne saurait rien faire de régulier."

[48] GP VI, 554. "Quant à ces substances immatérielles plastiques, je ne vois pas bien ni comment elles y suffisent, ni en quoi elles aient de l'avantage sur le Mécanisme. Il semble qu'il leur faudrait une intelligence toute divine pour former les organes des animaux avec la connaissance d'un architecte: mais si elles n'ont point cette connaissance architectonique, et s'il faut que Dieu y supplée et les dirige particulièrement, à quoi bon ces intelligences... ?" See Catherine Wilson, *Leibniz's Metaphysics* (Princeton, NJ: Princeton University Press, 1989), 160 *sq.*

important for Cudworth and Grew to emphasize the fact that some lower sorts of life are deprived of understanding and self-perception. As Cudworth writes:

> [There is a] certain Lower Life from the Animal, which acts *Regularly* and *Artificially*..., though it self do not know the Reason of what it does, nor is Master of that Wisdom according to which it acts, but only a Servant to it, and Drudging Executioner of the same.[49]

6. Conclusion

Contrary to Bayle's assumption, neither Grew nor Cudworth admit any autonomy of matter or something similar to the "substantial forms" of the Aristotelian tradition. It is precisely because they both want to refute materialist theses, presumably involved by this conception, that they promote two different conceptions of the notion of life, a generic one (life as something mental) and a specific one (life as vegetation), whose agreement is sometimes difficult to grasp. Indeed, despite the huge differences between the two authors, they both defend this double conception of life for apologetic reasons. On the one hand, the *mental* features of life understood as its higher capacities, sensation and intellection, are the best proof of the incorporeal nature of life, and this way, the best proof of the divine causation testified by vital phenomena. On the other hand, the *vegetative* sense of life, regarded as the lowest common denominator of the group of things or properties that are called 'vital,' is the best way to demonstrate both the presence of God everywhere throughout the whole universe and the unity of the creation. It is notably the adoption of these two ideas of life that paradoxically does not allow for a clear distinction between animate and inanimate bodies.

This ambiguity demonstrates the fecundity of the notion of life as a polemical tool. Nevertheless, in these conditions it is necessary to take into account the different dimensions of life: If it is a notion that can be determined by the phenomena described in the life sciences of that time, it keeps a strong theological dimension, which can partly modify its meaning.

Acknowledgments

I thank Justin Smith, Ohad Nachtomy, and Liat Lavi for their help.

[49] TIS, I, III, 37, §26, 172.

3

The Impossibility of a "Newton of the Blade of Grass" in Kant's Teleology

THOMAS TEUFEL

1. Introduction

As the last two decades of sustained historical scholarship on Immanuel Kant's mature philosophy of biology in the *Critique of the Power of Judgment*[1] have demonstrated, Kant's account of the role of teleological judgment in biological inquiry was both deeply informed by and had considerable influence on state-of-the-art work in the life sciences of his day. Critics have argued, however, that Kant had influence on figures like Blumenbach, Goethe, Schelling, and Kielmeyer primarily because his contemporaries misunderstood either Kant's position or its implications.[2] According to this line of criticism, Kant's critique of teleology never quite deserved the level of attention it received from practitioners and philosophers of biological science—then and now—because it in fact "delivered up a profound indictment of any biological discipline attempting to become a science."[3] That indictment is supposed to be epitomized by Kant's

[1] All references to Kant's works are to the 'Akademieausgabe' (*Ak.*), Immanuel Kant, *Gesammelte Schriften*, 29 vols., Königlich Preussische Akademie der Wissenschaften (Berlin: G. Reimer, 1900–). All references are preceded by standard abbreviations (*KU* for *Critique of the Power of Judgment*, *MAN* for *Metaphysical Foundations of Natural Science*). All translations are mine.

[2] Robert J. Richards, *The Romantic Conception of Life* (Chicago: Chicago University Press, 2002). John Zammito, "Teleology then and now: The question of Kant's relevance for contemporary controversies over function in biology," in *Kantian Teleology and the Biological Sciences*, ed. Joan Steigerwald, *Studies in History and Philosophy of Biological and Biomedical Sciences* 37.4 (2006): 748–770. Phillip R. Sloan, "Kant on the history of nature: The ambiguous heritage of the critical philosophy for natural history," in *Kantian Teleology and the Biological Sciences*, ed. Joan Steigerwald, *Studies in History and Philosophy of Biological and Biomedical Sciences* 37.4 (2006): 627–648.

[3] Richards, *Romantic Conception*, 229; Zammito, "Teleology then and now," 765–766.

famous denial of the possibility of a 'Newton of the Blade of Grass.' As Kant makes the point:

> .... one can brazenly say that it would be absurd for humans ... to hope that there may yet arise a Newton who could make comprehensible even the generation of a blade of grass according to natural laws that no intention had ordered (*KU*, §70, *Ak.* 5:400).

Rejecting the possibility of biology as a science on the grounds that biological explanations appeal to teleological paradigms that are not only ineliminable but, the criticism suggests, inconsistent with empirical causal analysis would, certainly in hindsight, appear to be a spectacularly bad call. Biology is alive and well as empirical causal science (and has long found its own Newton), regardless of continuing philosophical debates about the status and legitimacy of functional explanations. But did Kant really make the bad call in question? Is Kant's critique of teleology, in the words of one recent author, both "pointless in [biological] practice" and in denial about the truly "hard problems" in the philosophy of biology because it succumbs to transcendental-philosophical "scruples" about the grounds and warrant of our judgments of organic nature that could only haunt an 18th-century philosopher but need no longer trouble us?[4]

The utility of Kant's critique of teleology for the practice of biological science may certainly be doubted, if only because the relation between philosophical theory and scientific practice is complicated to begin with and hardly, if ever, clearly instrumental. Moreover, Kant's contemporaries may certainly have misunderstood Kant's position, if only because the outlines of that position are still very much the topic of debate today. By the same token, it appears that Kant's present-day critics are prepared, rather prematurely, to close the book on Kant's critique of teleology, as Kant arguably had a keener sense of what the truly 'hard problems' in the philosophy of biology are than those critics fathom.

In the background of my tale lies the common failure to give proper weight to the central development in Kant's philosophy of biology in the *Critique of the Power of Judgment*, his recognition of a genuinely a priori and necessary principle governing judgments of nature's purposiveness—the "transcendental concept of a purposiveness of nature" (*KU*, §V, *Ak.* 5: 184).[5] The charge that Kant thought it impossible to give a proper causal analysis of biological phenomena makes sense, if at all, only in the *absence* of that new transcendental principle.

[4] Zammito, "Teleology then and now," 766.

[5] On the difficult question of whether this principle actually *is* a genuinely a priori and necessary principle, see Thomas Teufel, "What does Kant mean by 'Power of Judgment' in his *Critique of the Power of Judgment*?" *Kantian Review* 17.2 (2012): 297–326.

The charge is predicated on the (as we will see) textually dubious idea that Kant, in the third *Critique*, recognizes a peculiar type of causal process in organic nature (roughly associated with organic self-reproduction), which we can readily experience but which, for Kant, allegedly remains conceptually inscrutable.[6] If Kant did offer a conceptually inscrutable causal process as the distinguishing feature of biological phenomena, then biology would indeed stand on (transcendental-) philosophically shakier ground than physics. Functional (i.e., teleological) explanations of biological phenomena would serve to pick up the explanatory slack but could aspire, at best, to heuristic status. They would be explanatory approximations to processes we do not fully understand because they cannot, ultimately, be understood.[7]

Kant's addition to critical philosophy of a transcendental principle of nature's purposiveness belies that account. Kant, in 1790, once again performs a Copernican Revolution in philosophy. This time arguing that certain central teleological dimensions of our cognition of the world are not grounded in empirical observation, but are instead manifestations of a priori and necessary (albeit in this case not categorial) features of self-conscious human experience. This is a significant departure from Kant's previous view, held as late as 1788, that "No human being can discern *a priori* that there must be purposes in nature" (*On the Use of Teleological Principles in Philosophy, Ak.* 8: 182). In 1790, armed with his new a priori principle of purposiveness, Kant insists on two related points. First, and contrary to a widely held misconception,[8] Kant believes that our judgments of biological purposiveness exhibit a form of necessity that goes beyond the merely subjective necessity he grants to heuristic conveniences (see sections 3 and 5). Second, and contrary to another widely held misconception, Kant believes that there are *no* perceptible-yet-inscrutable processes in empirical reality that might serve as the empirical touchstone for our judgments of biological purposiveness (whether those processes be forms of reciprocal, backward, whole-to-part, or some other form of putatively non-mechanistic causality).[9] For Kant, empirical reality—including *organic* empirical reality—remains

[6] Zammito, "Teleology then and now," 765.

[7] Of course, it is unclear how we can even consider them approximations (or how to analyze their heuristicity) when we cannot spell out just what they are approximations *to*.

[8] Peter McLaughlin, *Kant's Critique of Teleology in Biological Explanation* (Lewiston, NY: Edwin Mellen Press, 1990), 156. Henry E. Allison, "Kant's Antinomy of Teleological Judgment," in *Kant's Critique of the Power of Judgment: Critical Essays*, ed. Paul Guyer (Lanham, MD: Rowman & Littlefield, 2003), 219–236, 224. Marcel Quarfood, *Transcendental Idealism and the Organism*, Stockholm Studies in Philosophy 26 (Stockholm: Almqvist & Wiksell, 2004), 168. Hannah Ginsborg, "Kant's Biological Teleology and its Philosophical Significance," in *A companion to Kant*, ed. Graham Bird (Oxford, UK: Oxford University Press, 2008), 455–469, 460.

[9] The notion that, for Kant, there are such perceptible yet inscrutable processes has been most strongly defended in McLaughlin, *Kant's Critique of Teleology*, 173–174.

mechanistic (hence, *scrutable*) through and through, pace his denial of the possibility of a 'Newton of the Blade of Grass.'

There is, to be sure, something to that denial and to the idea that Kant advocates a kind of teleological inscrutability. Only, Kant's account is more radical—it functions at a deeper level—than typically appreciated. Kant, in the third *Critique*, considers the possibility of what we might somewhat ominously call 'dark' aspects of supersensible reality. These are aspects that must be considered utterly cognitively inert by the lights of the *Critique of Pure Reason* because they do not lend themselves to cognitive commerce with imagination and understanding at all. As a result, these aspects of supersensible reality cannot serve as a *ground* of (first-*Critique*) experience; for reasons that must, ex hypothesi, remain unknowable. Kant's new idea in the *Critique of the Power of Judgment* is that, if there were such aspects, they would nevertheless necessarily resonate with the reflecting power of judgment (now considered, for the first time in critical philosophy, as an upper cognitive faculty in its own right).[10] Specifically, such aspects would leave a unique (and ex hypothesi not *empirically* grounded) cognitive signature in the form of a reflecting judgment of purposiveness. Our apparent inability to judge biological phenomena except in functional terms—even though it is impossible ever to see a function or to observe a genuinely teleological process in nature[11]—is to be deemed an instance of just that signature judgment. That apparent inability thus serves as indirect confirmation not only that there are such mysterious aspects of supersensible reality but that an object that, *phenomenally*, appears as an organized being (on account of our empirically ungrounded teleological judging of it) is, *supersensibly*, a locale for them.[12]

On this reading, Kant's denial of the possibility of a 'Newton of the Blade of Grass' amounts to this: It is the recognition that any account of biological phenomena must explain the non-empirical, transcendental-logical sources of the phenomenal appearance of biological organization—in otherwise empirically fully causally determined entities. Kant's recognition of this circumstance has lost none of its urgency: Despite the best efforts of present-day teleological *realists*, a plausible non-reductive account of the phenomenal appearance of biological organization has not been forthcoming. Although it is possible to give non-reductive accounts of a variety of forms of metaphysical complexity, as well as of a range of non-standard forms of causality (including, if you flinch a bit, certain types of top-down causality), none of these accounts suffice to explain the genuine 'teleological extra' that teleological realism considers characteristic

[10] Teufel, "What does Kant mean by 'Power of Judgment'?"

[11] *KU*, §VII, *Ak*. 5: 189.

[12] Thomas Teufel, "What is the Problem of Teleology in Kant's *Critique of the Teleological Power of Judgment?*" *SATS: Northern European Journal of Philosophy* 12.2 (2011): 198–236, 231–234.

of self-reproducing organic systems.[13] Conversely, *deflationist* accounts that deny the possibility of a biologically grounded teleological extra and instead psychologize our teleological proclivities (by deeming function vocabulary in biology merely a useful explanatory device that has, over time, become habitual) typically fail to explain on what grounds the device first *attains* the utility that would help make it habitual. Kant's point in denying the possibility of a 'Newton of the Blade of Grass' is then, simply, that an account of the duly *non*-empirical and *non*-psychological grounds of the appearance of biological organization is something ex hypothesi beyond the reach of the empirical science of biology and, hence, something beyond the reach even of a veritable *Newton* of graminology.

If this background be granted, even if only for the sake of argument, then one may still wonder what might have prompted Kant to formulate the point in such potentially misleading language. After all, what the biological Newton is supposed to be incapable of explaining is the *generation* of a blade of grass—hence, presumably, the bi-parental origin of its seed, the seed's germination and growth into an adult plant, and so on. That is to say, while Kant's denial does *not* make explicit reference to the larger phenomenon of biological organization (as one might expect on my account), the processes he *does* reference are ones that we can—today—explain rather well. Accordingly, there remains a prima facie case to be made that Kant's denial of the possibility of a 'Newton of the Blade of Grass' is simply an instance of the inability of an 18th- century philosopher to imagine explanatory resources that would in due course become scientific routine.

In order to begin to understand Kant's actual point (and to motivate the notion that the account of a more radical teleological inscrutability just sketched may, in fact, be true), we need to take a closer look at Kant's denial of the possibility of a 'Newton of the Blade of Grass' in its proper context. The outsize influence often granted to that denial[14] is the result of a failure to appreciate fully its precise and limited role in Kant's teleological Dialectic in the second part of Kant's "Critique of the Teleological Power of Judgment." Specifically, Kant's denial of the possibility of a "Newton of the Blade of Grass" is a somewhat theatrical reaffirmation of the antithesis position in Kant's "Antinomy of the Power of Judgment" (*KU*, §70, *Ak.* 5: 385, caption) in that Dialectic. In order to understand Kant's denial, one must accordingly give due consideration to (i) the nature and structure of

[13] Thomas Teufel, "Wholes that Cause their Parts: Organic Self-Reproduction and the Reality of Biological Teleology," *Studies in History and Philosophy of Biological and Biomedical Sciences* 42.2 (2011): 252–260, 258–260.

[14] An influence that finds its expression in the deeply entrenched, but deeply problematic, view that Kant advocates the "mechanical inexplicability" of certain biological processes in nature. See, e.g., Hanna Ginsborg, "Two Kinds of Mechanical Inexplicability in Kant and Aristotle," *Journal of the History of Philosophy* 42.1 (2004): 33–65.

that antinomy (see section 2); (ii) the thing that the thesis position in Kant's antinomy affirms (see section 3); (iii) the thing that the antinomy's contravening antithesis position denies (see section 4); and (iv) Kant's proposed resolution to the antinomy, which further clarifies the meaning of both thesis and antithesis (see section 5). Without such contextualization, interpretations of Kant's denial of the possibility of a 'Newton of the Blade of Grass' must necessarily lack nuance and tend to remain inappropriately one-sided. Once the requisite context is provided, we find that Kant is very far indeed from denying the possibility of biology as a science. Instead—and fully consistent with the project of a *critique* of teleology—he denies that a comprehensive *philosophical* accounting for biological phenomena can do without at least some input from *transcendental* philosophy.

2. The Antinomy of the Power of Judgment

In §70 of the *Critique of the Power of Judgment,* Kant presents the central philosophical difficulty he believes that explanations of biological phenomena confront in the form of an 'antinomy of the power of judgment.' To express philosophical perplexities in the form of antinomical conflicts is a frequent strategy of Kant's. The general idea is to boil such perplexities down to a set of two contradictory propositions—a thesis and an antithesis—to whose conjunction reason is supposedly unavoidably committed, in order to then show (a) that the contradiction is resolvable (hence, only apparent) if (and only if) we give the propositions in question a transcendental-philosophical interpretation (as, accordingly, we must, lest reason be inconsistent with itself); and (b) that the *appearance* of a contradiction nevertheless remains (since, even after the resolution, our transcendental-philosophical interpretation is not reflected in the structure of the propositions in question). This strategy allows Kant to give sophisticated diagnoses of longstanding philosophical impasses all the while recommending his transcendental-philosophical middle ground as the only plausible non-dogmatic way out of them.

The teleological antinomy in the *Critique of the Power of Judgment* is somewhat unusual because both thesis and antithesis presuppose the results of the *Critique of Pure Reason.* It is a conflict that arises not from illicit dogmatic assumptions but on the grounds of critical orthodoxy itself. This naturally heightens the threat of a "contradiction in reason" (*KU,* §70, *Ak.* 5: 387)—i.e., the threat that the conflict might prove genuine (an unavoidable and irresolvable contradiction) as opposed to merely antinomical (an unavoidable appearance of a nevertheless resolvable contradiction)—since the conception of reason at issue is Kant's own critically chastened one. In order to resolve *this* kind of conflict, Kant cannot simply appeal to transcendental philosophy per se. He must appeal to a suitably beefed-up version of transcendental philosophy. Consistent with his central addition to the critical edifice in the *Critique of the Power of Judgment*, Kant, in his resolution of the teleological conflict, accordingly appeals to transcendental philosophy cum transcendental principle of purposiveness (see section 5).

The philosophical perplexity Kant addresses in his teleological antinomy concerns the legitimacy of function vocabulary for our explanation of biological phenomena. Specifically, it pits a form of deflationist reductivism, according to which all corporeal causation is mechanistic, against a form of non-reductive teleological realism, according to which function vocabulary tracks irreducibly teleological causal processes in nature. Kant narrowly frames the conflict in terms of the explanatory purchase of causal mechanism on natural (including biological) phenomena. Either causal mechanism's explanatory purchase on natural phenomena is (in a sense to be discussed) unlimited (thesis), thus confining function vocabulary to, at best, heuristic expediency; or causal mechanism's explanatory purchase on natural phenomena is (in a sense to be discussed) limited (antithesis), thus opening up the possibility that function vocabulary renders more fully intelligible those dimensions of the natural world mechanism is deemed to capture only incompletely.

Kant contends that in its explanation, specifically, of the causal *origin* of natural phenomena, our faculty of judgment "can set out from two maxims" (*KU*, §70, *Ak.* 5:386). The first maxim holds that, when investigating an object's causal origins, judging it mechanistically is always necessary:

> The first maxim of the power of judgment is the *thesis*: All generation of material things and their forms must be judged as possible in accordance with merely mechanistic laws (*KU*, §70, *Ak.* 5:387).

The second maxim, in contrast, envisions a limit to judging causal origins mechanistically:

> The second maxim is the *antithesis*: Some products of material nature cannot be judged as possible according to merely mechanistic laws (judging them requires an entirely different law of causality, namely that of final causes) (*KU*, §70, *Ak.* 5:387).

Kant's contention is that while reason "can set out" from one or the other of these maxims in our research into the causal origin of natural phenomena, reason is fundamentally committed to both. Simplifying a bit, reason is simultaneously committed to the principle that "*All* [natural origins] *must* be judged [merely mechanistically]" *and* to the principle that "*Some* [natural origins] *cannot* be judged [merely mechanistically]." It is important to note that the conflict at the heart of Kant's teleological antinomy is, accordingly, *not* a conflict between a mechanistic maxim and a teleological maxim. It is, instead, a conflict between competing interpretations of the nature and scope of the same underlying mechanistic maxim. I will henceforth call this underlying maxim—which enjoins us to judge natural origins merely mechanistically (whether *always* necessarily or *not* always necessarily)—the 'maxim of mechanistic judging.' To be sure, Kant's parenthetical appeal (in the antithesis position) to a separate teleological maxim offers a clue as to why reason might be committed to a limited interpretation of the maxim of mechanistic judging. Yet, despite its implications for the legitimacy of teleological explanations, Kant's 'teleological antinomy' is, in all strictness, a mechanistic antinomy.

The conception of mechanism at issue in this mechanistic antinomy is a broad one. Mechanism here is not restricted to billiard-ball or impact causality but instead governs action at a distance as much as emergent (part-to-whole), demergent (whole-to-part), and other forms of synchronic and diachronic causality. Kant's notion of mechanism is so accommodating because it is nothing but the category of causality under condition of corporeality. In the introduction to the third *Critique*, Kant explains that if we take the category of causality and substitute "the empirical concept of a body (as a moveable thing in space)" (*KU*, §V, *Ak.* 5: 181) for the ontological predicate of a "changeable substance…" (ibid.), we thereby transform the *transcendental* causal principle that "every event has a cause" (ibid.) into the *metaphysical* causal principle that "[all] alteration [of motile bodies in space] must have an *external* cause" (ibid., emphasis Kant's). If we presuppose corporeal motility, we can then infer "entirely *a priori*" (ibid.) the *externality* of all causal relations in the corporeal world. In the *Metaphysical Foundations of Natural Science* of 1786, Kant presents this metaphysical causal principle—that "Every change of matter has an external cause" (*MAN*, *Ak.* 4: 543)—as his "Second Law of Mechanics" (ibid.). In the *Critique of the Power of Judgment*, Kant streamlines this position and treats the metaphysical causal principle as identical with causal mechanism.

It must be emphasized that the principle at the heart of the antinomy in Kant's teleological Dialectic is *not* this objective (namely, *object*-determining) principle of causal mechanism. The principle whose divergent interpretations give rise to the antinomical conflict is the 'maxim of mechanistic judging' mentioned above: a subjective (namely, *judgment*-determining) principle or a "maxim of a reflecting power of judgment" (*KU*, §70, *Ak.* 5: 387). This circumstance has given rise to the widespread misconception that Kant's formulation of the teleological antinomy somehow reflects a demotion of universal causal mechanism to a rather more limited corporeal application.[15] But Kant's formulation

[15] McLaughlin, *Kant's Critique of Teleology*, 152–154.

of the antinomy involves no such thing.[16] The maxim of mechanistic judging is a technical-practical principle of cognitive conduct by means of which the reflecting power of judgment (whose job is judgment-determination) instructs the determining power of judgment (whose job is object-determination) how to go about its object-determining business.[17] Roughly, the reflecting power of judgment here instructs the determining power of judgment: 'If you want to judge causal origins in nature, you must judge them mechanistically.' Assuming that a base condition of agent rationality is satisfied, the rationale behind this instruction is then the objective validity of the metaphysical causal principle. We must judge causal origins in nature mechanistically because, according to the principle of causal mechanism, all alteration of bodies in space must have an external (i.e., mechanistic) cause. We will see shortly that the source of the antinomy is that we nevertheless seem inherently torn as to how stringent this judgment-determining instruction is to be interpreted. The point I wish to make here is that appeal to the principle of causal mechanism in a merely subjective, judgment-determining maxim does not at all reflect a somehow diminished status for the principle of causal mechanism. Indeed, there would be no instrumentally binding, judgment-determining 'maxim of mechanistic judging' if an objectively valid 'principle of causal mechanism'—from which alone such a maxim can be rationally derived—were not in full force.[18] Or, as Kant puts it, given the end of judging causal origins in nature and given the principle of causal mechanism governing those origins, the maxim of mechanistic judging is "provided to [judgment] by the mere understanding a priori" (*KU*, §70, *Ak.* 5: 386).

3. Thesis: The *Possibility* of a 'Newton of the Blade of Grass'

It follows from the above that for any cognitive agent who pursues the end of judging causal origins in nature (whether scientist or police investigator), the assumption of agent rationality, in combination with the objective validity, for corporeal nature, of the metaphysical causal principle, straightforwardly entails that the agent's judgment must be constrained by the thesis interpretation of the maxim of mechanistic judging: If all generation of material things and their forms really is mechanistic, then "All generation of material things and their

[16] Teufel, "What is the Problem of Teleology?" 221–226.

[17] Teufel, "What is the Problem of Teleology?" 230f.

[18] Ken Westphal has demonstrated that Kant is unable to establish this principle on the grounds of his transcendental and metaphysical analyses in the *Critique of Pure Reason* and in the *Metaphysical Foundations of Natural Science* alone. Ken Westphal, *Kant's Transcendental Proof of Realism* (Cambridge, UK: Cambridge University Press, 2004), 214–223.

forms must be judged as possible in accordance with merely mechanistic laws" (*KU*, §70, *Ak.* 5: 387). If we are interested in arriving at a genuine explanation of causal origins in nature, then those origins must not be judged in accordance with non-mechanistic forms of causality—because there are none.

It is easy to see that Kant's apodeictic formulation of the thesis interpretation of the maxim of mechanistic judging ("*All*... *must*") thus expresses a technically-practically binding form of necessity. Judging in accordance with merely mechanistic laws is not simply desirable or recommended but universally necessary for rational investigators who genuinely seek to explain causal processes in nature. Note that, to the extent that biological phenomena are natural phenomena, there is then no in-principle barrier to the notion "that there may yet arise a Newton who could make comprehensible even the generation of a blade of grass according to natural laws that no intention had ordered" (*KU*, §70, *Ak.* 5: 400). According to the *thesis* position, the emergence of a 'Newton of the Blade of Grass' is very much a live possibility. Indeed, it is rather to be expected, given reason's tendency to unify its cognitions—if only nature cooperates and proves amenable to our unifications.

Accordingly, the view that—qua subjective (judgment-determining) principles—the regulative maxims expressed in thesis and antithesis can only exhibit a subjective form of necessity, recommending how we *must seek to (but need not) judge*, as opposed to dictating how we *must judge*, is misguided. Regulativity, for Kant, does not entail heuristicity.[19] While all heuristic (merely subjectively necessary) principles are regulative (judgment-determining) principles, not all regulative principles are heuristic. This is evident, moreover, from the status of the most important regulative principle in Kant's *Critique of the Power of Judgment*, his new principle of purposiveness. This principle is merely "a subjective principle (maxim)" (*KU*, §V, *Ak.* 5: 184) of cognitive conduct that governs the reflecting power of judgment. And yet, Kant's point in offering a *Critique* of the reflecting power of judgment and a transcendental deduction of its governing principle is to show that this subjective maxim holds with unmitigated transcendental necessity.[20]

If a principle's regulativity thus does not entail its heuristicity (and is, instead, consistent with its technical-practical or even transcendental necessity), then one of the traditional problems in interpreting Kant's antinomy is readily solved.[21] It is very hard indeed to see how divergent heuristic maxims of judgment—instructing how we *must seek to (but need not) judge*—can give rise to an unavoidable appearance of a contradiction. After all, the regulative

[19] Teufel, "What is the Problem of Teleology?" 209–213.

[20] Teufel, "What does Kant mean by 'Power of Judgment'?" 306.

[21] Teufel, "What is the Problem of Teleology?" 200, 216f.

(judgment-determining) demands always to seek to judge mechanistically and sometimes to seek to judge non-mechanistically are fully compatible—precisely because neither dictates the (thus) merely pragmatically recommended and not universally necessary judgment in question (indeed, the demands are complementary because the latter heuristic strategy is poised to kick in if and when the former runs out of steam). In contrast, it is not hard at all to see the contradiction between the regulative (judgment-determining) demand always to judge mechanistically and the regulative demand sometimes not to.

4. Antithesis: The *Impossibility* of a 'Newton of the Blade of Grass'

Before we can understand the nature of this contradiction as merely antinomical—i.e., as the unavoidable appearance of a nevertheless resolvable contradiction—we must note what is perhaps the most surprising feature of Kant's antinomy, and the central element for any measured understanding of Kant's denial of the possibility of a 'Newton of the Blade of Grass.' It is the rarely appreciated fact that Kant's antithesis position unreservedly *endorses* the universal necessity of judging causal origins in nature in accordance with merely mechanistic laws. To be sure, the sticking point in Kant's antinomy is the proper interpretation of the nature and scope of the underlying maxim of mechanistic judging. But, as it turns out, the *universal necessity* of that maxim is simply *not* at issue in Kant's antinomy—it is, instead, presupposed throughout. Kant is quite unambiguous about this in his explication of the antithesis position, immediately following the presentation of the antinomy. Kant notes that when (for reasons he does not fully disclose at this point) we deem appeal to non-mechanistic (specifically teleological) principles necessary for judging the causal origin of certain "forms of nature" (e.g., organisms), this does not mean that "those forms *would not be possible* in accordance with the mechanism of nature" (*KU*, §70, *Ak.* 5: 388, my emphasis). Nor does judging causal origins in accordance with the antithesis interpretation of the maxim of mechanistic judging mean that "reflection in accordance with the first maxim *is . . . thereby suspended*" (ibid., my emphasis). In other words, if the occasional necessity of non-mechanistic (specifically teleological) judging does not suspend "reflection in accordance with the first maxim," and if the latter is a universally necessary constraint on judging causal origins in nature (see section 3), then the occasional necessity of judging causal origins in nature teleologically does not suspend the necessity, let alone the possibility or desirability, of judging those origins mechanistically. Given Kant's steadfast commitment to the objective validity of the metaphysical causal principle, this is, moreover, the only consistent position he can take.

But what—if not the universal necessity of judging causal origins mechanistically—does the antithesis interpretation of the maxim of mechanistic judging reject? What—if not that universal necessity—is the antinomical conflict about? Here, too, Kant's text is straightforward. Kant's formulation of both thesis and antithesis makes clear that the pertinent issue in the antinomy is not the necessity of judging causal origins in nature mechanistically but the necessity of judging causal origins in nature *merely* mechanistically. The thesis contends that "All generation of material things and their forms must be judged as possible in accordance with *merely* mechanistic laws" (*KU*, §70, *Ak.* 5: 387, my emphasis). The antithesis contends that the generation of some material things and their forms "cannot be judged as possible according to *merely* mechanistic laws" (ibid.). What is at issue in the antinomy is thus not the necessity but the exclusivity—and, hence, the *sufficiency*—of mechanistic judging. Although both sides of the conflict agree that judging causal origins in nature mechanistically is a universally necessary constraint on a rational investigator's explanatory practices, the thesis contends—and the antithesis disputes—that this is therefore the *only* necessary constraint on judging causal origins in nature. The antithesis position curbs the scope of the maxim of mechanistic judging by envisioning the occasional co-necessity of other, specifically teleological forms of judging.

The conflicting positions in Kant's antinomy must, then, be understood as follows. The thesis contends that judging mechanistically is both *universally necessary* and *universally sufficient* for the rational investigation of causal origins in nature: A rational investigation of causal origins in nature must consider *all and only* mechanistic causes. The antithesis contends that judging mechanistically is *universally necessary* but not therefore *universally sufficient* for the rational investigation of causal origins in nature: A rational investigation of causal origins in nature must occasionally appeal to non-mechanistic (specifically teleological) causes as well.

As noted, Kant's contention is that, by the very nature of our cognitive faculties, we are simultaneously committed to both of these interpretations of the maxim of mechanistic judging. We will see momentarily why this should be so and how Kant proposes to resolve the conflict, exposing it in one sweeping maneuver as unavoidable and, yet, merely apparent. For now, an answer to the original question pursued in this chapter begins to suggest itself. When Kant denies the possibility that there may yet arise a 'Newton of the Blade of Grass,' he does not deny the necessity, let alone the possibility, of a thoroughgoing mechanistic explanation of organic phenomena. A fortiori, he does not deny the possibility of a scientist who might yet offer such an explanation (although it would appear that the Newton whose possibility Kant thus does envision is more akin to Watson and Crick than Darwin and Wallace). What Kant *does* deny is that the mechanistic explanations such a Newton (and, a fortiori, a mature science of biology) can provide, no matter how complete, could possibly offer an exhaustive account of the organic phenomena under investigation. This focus on the universal sufficiency of mechanistic judging is very much reflected in the text itself. What Kant diagnoses as "absurd" in his denial of the possibility of a 'Newton of the Blade of Grass' is not the pursuit of thoroughgoing mechanistic explanations of organic phenomena per se. What Kant diagnoses as absurd is the notion that this pursuit should ever yield an "adequate" (*KU*, §75, *Ak.* 5: 400)—hence, sufficient—account of these phenomena.

5. Resolution: The Transcendental Principle of Purposiveness

We have, however, reached an impasse. The thesis interpretation of the maxim of mechanistic judging is fully consistent with Kant's commitment to the objective validity of the metaphysical causal principle. If corporeal causality is all and only mechanistic, then our judging of corporeal causality must appeal to all and only mechanistic causes. The antithesis interpretation, in contrast—since likewise committed—seems to border on incoherence, thereby threatening Kant's formulation of the antinomy at large. Either appeal to non-mechanistic causes proves occasionally genuinely necessary (whether technically-practically or in some other sense of necessity). But then the metaphysical causal principle, which ostensibly rules out the possibility of such causes (and hence the necessity to ever have to appeal to such causes in our explanation of natural phenomena), would fail to be universal; or else appeal to non-mechanistic causes is not ever truly necessary (whether technically-practically or in some other sense of necessity). But then the notion of a contradiction with the thesis interpretation of the maxim of mechanistic judging would be without merit, since non-mechanistic judging would always remain ultimately optional. It seems that either the thesis position or the antithesis position—and with it the antinomy at large—must collapse, or else critically chastened reason itself would prove incoherent.

The difficulty here stems from the fact that the occasional necessity of judging teleologically, which prompts the antithesis, cannot derive from any type of causality that characterizes empirical reality and yet must exhibit more than a merely subjective (e.g., psychological) form of bindingness. This combination of factors is impossible to square on first-*Critique* grounds. The difficulty resolves rather quickly, however, when considering that teleological judgments (and, generally, judgments of purposiveness) in the third *Critique* derive from Kant's new 'transcendental principle of purposiveness.' Although Kant's claims surrounding the nature and legitimacy of this principle are difficult and, to this day, not very well understood,[22] the general structure of Kant's resolution of the conflict between thesis and antithesis now comes into view.

The principle of judgment that motivates the antithesis position (i.e., the position that mechanistic judging, though always necessary, is occasionally insufficient for a thoroughgoing account of certain products of material nature) is the teleological maxim parenthetically attached to Kant's formulation of the antithesis. It holds that "judging [some products of material nature] requires an entirely different law of causality, namely that of final causes" (*KU*, §70, *Ak.* 5: 387). This teleological maxim registers a different type of necessity than the maxim of mechanistic judging (which announces the technical-practical necessity of judging causal origins mechanistically). The 'requirement' of having recourse to a law of final causes that Kant's teleological maxim lays down concerns Kant's transcendental principle of purposiveness—there is no other source for the putative necessity of teleological judging in the third *Critique*. But this transcendental principle of purposiveness is a regulative, judgment-determining principle, not a constitutive, object-determining principle. It is thus not the teleological equivalent of the metaphysical causal principle. Given the universal purchase of the metaphysical causal principle on material nature and the consequent absence of corporeal teleological causes, there can then be no instrumental reason—grounded in objective principle—to judge organisms teleologically. Not mediated by instrumental rationality, the necessity of judging certain products in nature teleologically can then only be in some sense immediately reflective—a manifestation in phenomenal consciousness—of the otherwise sub-personal transcendental necessity of the principle of purposiveness.[23]

If this is so, then these necessary teleological judgments have neither empirical ground nor objective import. Kant makes this explicit for a priori judgments of aesthetic purposiveness—although the point holds with perfect generality for all judgments of the reflecting power of judgment (whether a priori aesthetic or necessarily teleological): "the purposiveness of a thing, insofar as it is represented in perception is also not a property of the object itself (for such a thing cannot be observed)" (*KU*, §VII, *Ak.* 5: 189). Necessary judgments of purposiveness, though inescapably made in the presence of certain products of nature (as well as certain products of art) and thus inextricably linked with our phenomenal awareness of those products, are not reflective of any objective, empirically discernible purposive features in them. Contra Paley, we do not, because we cannot, directly perceive purposes.

By thus tying the apparent inescapability of our appeal to function vocabulary in our judging of organic phenomena to the judgment-determining

[22] Teufel, "What does Kant mean by 'Power of Judgment'?" 324f.

[23] Teufel, "What is the Problem of Teleology?" 231–234.

transcendental principle of purposiveness, Kant at once suggests—in the sweeping maneuver mentioned before—why we cannot simply kick our teleological habit *and* why this inability nevertheless does not conflict with our rational mechanistic obligations. Teleological judgments of organic nature, when and where we do make them,[24] are, in their own unusual way and by the nature of our cognitive faculties, as necessary (if not as universally necessary) as their rationally required mechanistic counterparts. The conflict between the universal necessity and universal sufficiency of judging products of nature mechanistically (thesis) and the occasional insufficiency of judging products of nature mechanistically (antithesis) is thus unavoidable. But once we locate the source of the teleological judgments that account for that occasional insufficiency in the transcendental principle of purposiveness, we immediately see that the peculiar necessity these judgments derive from that principle in fact leaves both the universal necessity *and* the universal sufficiency of judging corporeal nature mechanistically—fully unscathed. A comprehensive account of the *corporeal reality* of organic beings must appeal to *all and only* mechanistic causes. A comprehensive account of our *phenomenal awareness* of organic beings will, in contrast, have to have recourse to critical philosophy cum transcendental principle of purposiveness. Kant's denial of the possibility of a 'Newton of the Blade of Grass' is, then, not a rejection of the possibility of biology as a science. After all, a comprehensive account of the corporeal reality of organic beings—not of our phenomenal awareness of them—is what the science of biology and its Newton care about.

[24] The question of just what, if not experience, could possibly 'prompt' these necessary yet nevertheless particular judgments is vexing. The answer must lie in the peculiar interplay between the supersensible ground of experience and the cognitive role of the transcendental principle of purposiveness. Teufel, "What is the Problem of Teleology?" 234.

PART TWO

THE STRUCTURE OF LIVING BEINGS

4

Fabricius's Galeno-Aristotelian Teleomechanics of Muscle

PETER DISTELZWEIG

1. Introduction

Hieronymus Fabricius ab Aquapendente (Girolamo Farici, 1533?–1619) has attracted the attention of historians of science as an important contributor to Renaissance anatomy. For example, Andrew Cunningham has done much to help us understand Fabricius's approach to anatomical research (what Cunningham calls his "Aristotle Project").[1] More recently, Cynthia Klestinec has provided an insightful discussion of Fabricius's career, focusing on pedagogy, private and public anatomical venues, and the experience of his students at Padua.[2] In addition, because of his use of mathematical mechanics in his work on muscle anatomy,

[1] See Andrew *Cunningham*, "Fabricius and the 'Aristotle project' in anatomical teaching and research at Padua," in *The Medical Renaissance of the Sixteenth Century*, eds. Andrew Wear, Roger French, and Iain Lonie (Cambridge, UK: Cambridge University Press, 1985), 195–222; and *The Anatomical Renaissance: Resurrection of the Anatomical Projects of the Ancients* (Aldershot, UK: Scolar Press, 1997). For a general introduction to Fabricius's life, career, and publications, see Howard Adelmann, trans., *The Embryological Treatises of Hieronymus Fabricius of Aquapendente*, 2 vols. (Ithaca, NY: Cornell University Press, 1942), 3–35; and Maurizio Rippa Bonati, "Girolamo Fabrici d'Acquapendente: per una bio-crono-bibliografia," in *Il Teatro dei Corpi:Le* Pitture Colorate D'Anatomia *di Girolamo Fabrici d'Acquapendente*, eds. Maurizio Rippa Bonati and José Pardo-Tomás (Milano, ES: Mediamed Edizioni Scientifiche Srl., 2004), 268–277. For a helpful, more general discussion of humanistic medicine at Padua, including anatomy, see Jerome Bylebyl, "The School of Padua: humanistic medicine in the sixteenth century," in *Health, medicine and mortality in the sixteenth century*, ed. Charles Webster (Cambridge, UK: Cambridge University Press, 1979) 335–370.

[2] Cynthia Klestinec, *Theaters of Anatomy: Students, Teachers and Traditions of Dissection in Renaissance Venice* (Baltimore, MD: Johns Hopkins University Press, 2011). For a discussion of Fabricius's approach in relation to that of his most famous student, William Harvey, see Roger French, *William Harvey's Natural Philosophy* (Cambridge, UK: Cambridge University Press, 1994), chapters 2 and 3 (esp. 64–68); Andrew Cunningham, "Fabrici and Harvey," in *Harvey e Padova: Atti del convegno celebrativo del quarto centenario della laurea di William Harvey (Padova, 21–22 novembre*

Fabricius has received attention from Ugo Baldini, in his study of treatments of animal motion prior to Giovanni Borelli's 1680/1681 *De motu locali animalium.*[3] Julian Jaynes also briefly discusses Fabricius and his use of mechanics in an article on the problem of animal motion in the 17th century.[4] However, general treatments like Cunningham's give little attention specifically to his work on muscle anatomy and his use of mechanics there, and neither Jaynes nor Baldini locates Fabricius's work on these topics within his larger anatomical project. The result is a misunderstanding of Fabricius's use of mechanics by Baldini and Jaynes, on the one hand, and an impoverished understanding of the extent of Fabricius's creative development and extension of Aristotelian and Galenic resources, on the other.

In this chapter, I argue that Fabricius attempts to integrate mathematical mechanics into teleological explanations of muscle anatomy characteristic of his general anatomical project. These explanations are structured by a Galenic distinction between the composition of a part, its action (*actio*), and its use (*utilitas*), and by an Aristotelian emphasis on providing systematic teleological explanations of the presence and variation in anatomical detail of related parts across species.[5] This analysis of Fabricius's use of mechanics stands in contrast to that suggested by Jaynes and Baldini. Both scholars suggest that Fabricius provides an explanation of animal locomotion in terms of a descriptive mechanics of the animal. My analysis shows that Fabricius is providing explanations of the *parts* of animals, not of their *locomotion*. The principles of mechanics, I argue, play the role in Fabricius's project that matter theory generally does: They undergird the hypothetical necessity of certain anatomical features of a part, given its function. Finally, I argue that Fabricius's use of mechanics is Aristotelian: The pseudo-Aristotelian[6] *Quaestiones Mechanicae* serve as his primary reference

2002), eds. Giuseppe Ongaro, Maurizio Rippa Bonati, and Gaetano Thiene (Padova, IT: Universita degli Studi di Padova, 2006), 129–149; and Peter Distelzweig, *"Meam de motu & usu cordis, & ciruitu sanguinis sententiam*: Teleology in William Harvey's *De Motu Cordis," Gesnerus—Swiss Journal of the History of Medicine and Science* (forthcoming); and "Descartes' Teleo-mechanics in Medical Context: Approaches to Integrating Mechanics and Teleology in Hieronymus Fabricius ab Aquapendente, William Harvey, and René Descartes" (Ph.D. diss., University of Pittsburgh, 2013).

[3] Ugo Baldini, "Animal Motion before Borelli, 1600–1680," in *Marcello Malpighi: Anatomist and Physician*, ed. Domenico Bertoloni Meli (Firenze, IT: Olschki, 1997), 193–246.

[4] Julian Jaynes, "The Problem of Animate Motion in the Seventeenth Century," *Journal of the History of Ideas* 31.2 (1970): 219–234.

[5] Although I have benefited from Cunningham's work, I agree with Siraisi that his preoccupation with identifying Aristotelian aspects of Fabricius's work led him to miss the fundamental importance to Fabricius of certain Galenic concepts, including the action/use distinction; see Nancy Siraisi, "*Historia, actio, utilitas*: Fabrici e le scienze della vita nel Cinquecento," in *Il Teatro dei Corpi:Le* Pitture Colorate D'Anatomia *di Girolamo Fabrici d'Acquapendente*, eds. Maurizio Rippa Bonati and José Pardo-Tomás (Milano, ES: Mediamed Edizioni Scientifiche Srl., 2004), 63–73.

[6] Fabricius attributes the work to Aristotle.

point, and he thinks of mechanics as an Aristotelian subordinate mathematical science that (a) identifies quantitative, causally interrelated features present in the natural world, but (b) treats them as a small subset of the features of natural things. His is a realist and non-reductionist application of mechanics to nature.

Fabricius's Galeno-Aristotelian "teleomechanics" provides an important example of the vibrancy and creativity of late Aristotelianisms in a medical context.[7] It challenges or at least complicates standard accounts of a 17th-century triumph of mechanics *over* Aristotelianism.

2. Fabricius's Mechanics of Animal Motion?

Fabricius's use of mechanics is most conspicuous in his discussion of the *utilitates* of muscles. It is here that we encounter Fabricius employing a number of more and less abstract diagrams in his analysis of muscles in terms of levers. Here Fabricius refers to the *Quaestiones Mechanicae,* distinguishes between the views of Galen and those of "mathematicians or practitioners of mechanics," and contrasts "natural" and "mathematical or mechanical" causes of anatomical facts.[8]

In his sweeping article on animate motion in the 17th century, Julian Jaynes begins with a brief treatment of Fabricius, focusing on his efforts in *De motu locali animalium, secundum totum* to connect different kinds of animal progression with differences in the animals' environments. Jaynes says, "...Fabricius separated out the biological problems of animal motion. He failed to find a solution, but succeeded in correlating animal motion with environment in a remarkably prescient way."[9] Of Fabricius's use of mechanics, Jaynes says only that "[n]ot having the advantage of Galileo's *Della Scienza Mecanica* which did not appear until 1634, Fabricius's mechanics of these motions only began what Borelli was to complete at the end of the century."[10] Jaynes characterizes these later efforts of Borelli as in part solving "much of the problem Fabricius and the

[7] In this way it complements previous work on Renaissance and early modern Aristotelianisms; see, e.g., Charles Schmitt, *Aristotle and the Renaissance* (Cambridge, MA: Harvard University Press, 1983); and Christia Mercer, "The Vitality and Importance of Early Modern Aristotelianism," in *The Rise Of Modern Philosophy: The Tension Between the New and Traditional Philosophies from Machiavelli to Leibniz,* ed. Tom Sorell (Oxford, UK: Oxford University Press, 1993), 33–67.

[8] For example, "*An ergo de vecte bene resondet Arist. q. 3. Mechanica*?" in Hieronymus Fabricius ab Aquapendente, *Opera omnia anatomica & physiologica* (Lipsiae, DE: Sumptibus Johannis Friderici Gleditschii, Excudebat Christianus Goezius, 1687), 419; and "*juxta Aristotelem in Q. 29. Mechanica,*" 420. (All references to Fabricius's work are to this edition, unless noted. All translations are mine, unless noted.) For the contrast between Galen and the mathematicians and between natural and mathematical cum mechanical causes, see 416–417.

[9] Jaynes, "Animate motion," 232.

[10] Jaynes, "Animate motion," 221.

Aristotelian writings had opened.... Borelli showed the bones to be true levers, that the length of the limb, the distance to the muscle, or differences in the center of articulation all influence the force necessary to make the movement."[11] Jaynes suggests that Borelli provides a particular kind of explanation of animal motion "found in the descriptive mechanics of the organism treated as a physical system."[12] In Jaynes's story, Fabricius is a "proto-Borelli," attempting to explain animal motion by providing a descriptive mechanics of the animal.

Fabricius's use of mechanics receives more detailed consideration from Ugo Baldini, who traces the origin of discussions of animal motion in the 17th century to two Aristotelian texts:

> [Aristotle] provided a phenomenology of animal motions in the opuscule which Greek tradition entitled περί πορείας ξώων and Middle Ages *De animalium incessu* (*De animalium progressu, De animalium gressu*). Then he provided a sort of psychophysiology of spontaneous motions in the περί ξώων κινήεως (*De motione animalium* or *De motu animalium*). The two works are largely unrelated: the first is mostly descriptive; the second lacks any mechanical character, and almost nowhere refers to motions or single structures producing them.[13]

Baldini is concerned about tracing how certain features of Aristotle's accounts in *De motu animalium* and *De incessu animalium* fare in each thinker he examines. He is also concerned with comparing each thinker's efforts to Borelli's. Baldini distinguishes a number of different facets of 17th-century analyses of animal motion and places Fabricius at the beginning of his account of one of them: efforts to analyze movements of parts of the body in terms of levers. He is preoccupied with the extent to which Fabricius treats the limbs as levers and how much of the analysis to be found in Borelli's later work is already present in Fabricius.[14] Thus Baldini also treats Fabricius as a "proto-Borelli" and (at least implicitly) shares the view that Fabricius's use of mechanics is at the service of providing explanations of animal motions by analyzing anatomical structures in mechanical terms.

If one is preoccupied with the rise of mathematical mechanics and the mechanical philosophy—and iatromechanism, in particular—it is perhaps natural to assimilate Fabricius's use of mechanics to a familiar picture of a

[11] Jaynes, "Animate motion," 232.

[12] Jaynes, "Animate motion," 233.

[13] Baldini, "Animal motion," 194.

[14] For example, in his summary remarks closing his section on Fabricius, Baldini writes, "A long distance has been travelled from Aristotle's model, but an equally long one still separated Fabrici [Fabricius] from Borelli's approach" (Baldini, "Animal motion," 208).

17th-century triumph of mechanism over Aristotelianism and its faculties and final causes. However, there is reason to be suspicious of this assimilation. Jaynes and Baldini both fail to locate Fabricius's use of mechanics in his larger anatomical project. Indeed, Jaynes is almost completely silent on Fabricius's intellectual context—save connecting his work with Galileo's in a vague and speculative way.[15] Although Baldini has improved the interpretation by connecting Fabricius's work to Aristotle's treatises on animal locomotion, he mischaracterizes the project of the *De incessu animalium* in an important way. Although this short treatise does provide a description of the limbs and motions involved in progression, its project is not descriptive but explanatory. Aristotle is providing final causal explanations of the presence of locomotive organs and their variations. This fact, reflected throughout the short work, is clear enough from its opening lines:

> We have now to consider the parts which are useful to animals for movement in place; first, why each part is such as it is and to what end they possess them; and second, the differences between these parts both in one and the same creature, and again by comparison of the parts of creatures of different species with one another.[16]

After providing a list of features of the locomotive parts (and related facts about how they move) Aristotle concludes, "[w]e have to examine the reasons for all these facts, and others cognate to them; that the facts are such is clear from our natural history, we have now to ask reasons for the facts" (704b8–11). Aristotle's project is not descriptive but explanatory and provides (teleological) explanations of animal *parts*, not animal *motions*. Might not Fabricius's texts, a part of what Cunningham has called Fabricius's "Aristotle project," share similar goals?

Another reason to wonder whether Baldini succeeds in carving Fabricius's work at the joints is the difficulty with which he connects the structure of Fabricius's texts to his own categorization of the topics and issues under consideration:

> Firstly [Fabricius] resolved the Aristotelian duality of dynamical and phenomenological into more specialized subjects. Physiological themes he mostly studied in *De musculis*; anatomical ones partly in the same work, partly in *De ossium articulis*; biomechanical aspects mainly in the third part of *De musculis*; phenomenology of animal motion in *De motu locali animalium secundum totum*.[17]

[15] See, e.g., Jaynes, "Animate motion," 220.

[16] *De incessu animalium* 704a4–8. The translations of Aristotle (including passages from the Pseudo-Aristotelian *Quaestiones Mechanicae*) are taken from Jonathan Barnes, ed., *The Complete Works of Aristotle* (2 vols.) (Princeton, NJ: Princeton University Press, 1984).

[17] Baldini, "Animal motion," 205.

Clearly, Baldini's conceptualization of the topics covered differs from the understanding shaping Fabricius's publications. This same point is reflected also in Baldini's silence regarding the near ubiquitous division of Fabricius's works into three sections, the first on structure or fabric of a part, the second on its action, and the third on its *utilitates*.[18] Baldini does not consider or explain why Fabricius's prominent use of mechanics appears in the section on the *utilitates* of muscle and not in the sections on its *fabrica* or *actio*. A satisfactory interpretation should explain this fact and place Fabricius's use of mechanics in the context of his work on muscles and his general anatomical project.

3. "Teleomechanics" in Fabricius's Galeno-Aristotelian Anatomy of Muscle

Fabricius published his works on muscles and joints together. He describes their scope in the opening sentences:

> We will now treat first concerning the fabric of the muscles, and then also concerning the joints of the bones. These are the principle and proximate instruments of the locomotion of the animal, and locomotion is judged the most perfect action of the animal. . . .[19] (383)

This identification of parts according to the action they perform is characteristic of what could be called Fabricius's *systems-based* approach to anatomical research. He did not publish a universal anatomy like Vesalius's *De Humani Corporis Fabrica* (1543) or Laurentius's *Historia Anitomica Humani Corporis* (1600). Rather, he published smaller works focused on particular physiological systems; each work focused on groups of parts at the service of some particular animal function.[20] An examination of Fabricius's publications makes this clear. For example, Fabricius published works like *De locutione & ejus instrumentis* and *De respiratione & ejus instrumentis*, which explicitly pick out the parts to be studied as those that are the instruments of certain activities. In addition, works like

[18] I return to this tripartite structure later.

[19] By locomotion, Fabricius means not only "progression" (when the whole animal changes location), but any locomotion exhibited by any part. This includes the motions involved in functions as diverse as eating, looking, and expelling excrement.

[20] That said, Fabricius did intend all the works to be seen as part of one project. He suggests that students could eventually bind all the individual tracts together. See the dedication to Hieronymus Fabricius ab Aquapendente, *De venarum ostiolis* (Patavii, IT: ex Typographia Laurentij Pasquati, 1603). A translation can be found in K. J. Franklin, trans., *De Venarum Ostiolis 1603 of Hieronymus Fabricius of Aquapendente (1533?–1619)* (Springfield, IL: Charles C. Thomas, 1933).

De larynge, vocis instrumento; *De oculo, visus organo*; and *De aure, auditus organo* also identify the anatomical parts to be studied precisely in terms of the function they serve. Similarly, Fabricius published *De gula*; *De ventriculo*; *De omento*; *De varietate ventriculorum*; *De intestinis*; and *De mesenterio* together in one volume. The opening lines of this text exhibit Fabricius's system-based approach:

> We now treat the stomach, and we will join with it also those things which are connected and go with it—that is, the intestines and the esophagus—and, for the same reason, the mesenterium, the bowels, and the muscles of the anus and of the abdomen, as much as these all are parts that, as a kind of chain, at the same time are conjoined and are aided by one another's roles. (99)

Besides exemplifying his system-based approach, Fabricius's work on muscles exhibits the ubiquitous tripartite structure characteristic of his works, devoting a section each to the *historia* of muscle, the treatment of its action, and the discussion of its *utilitates*. In his first publication, Fabricius provides a brief characterization of these three aspects of his anatomical project:

> *Historia* sets out the features present in the organ—temperaments, features following from temperaments, and accidents. . . . *Action* is each organ's office [*munus*] or function [*functio*]. *Utilitas* is nothing other than *that for the sake of which* that very action and also the features in the organ are present.[21]

The concepts of *actio* and *utilitas* and the distinction between them have their origin in Galen. In the "*Epode*" (i.e., Book 17) of *De usu partium*, a treatise devoted to studying the *utilitates* of the parts, Galen writes:

> Now the action [*energeia*] of a part differs from its usefulness [*chreias*], as I have said before, because action is active motion and usefulness is the same as what is commonly called utility [*euchrestia*].[22]

The action of a part, for Galen, is the active motion (in the broad sense of "change") by which it contributes to the functioning of the animal. For example, vision is said to be the action of the eye, and grasping that of the hand. Usefulness

[21] Hieronymus Fabricius ab Aquapendente, *De visione. De voce. De auditu* (Venetiis, IT: Per Franciscum bolzettam, 1600), second page of dedication to *De visione* (emphasis added).

[22] Margaret May, trans., *Galen: On the usefulness of the parts of the Body* (Ithaca, NY: Cornell University Press, 1968), 724.

(what Fabricius translates *utilitas*), in turn, is divided into the *utilitates* of actions and parts.

> Hence the usefulness of first importance to animals is that which is derived from actions and the second is that from the parts; for there is no part which we desire for its own sake, and a part deprived of its action would be so superfluous that we should cut it off rather than wish to keep it.[23]

The *utilitates* of an *action* are the ways in which the action contributes to the life of the animal. The *utilitates* of a *part*, in turn, are the ways in which the features and components of that part contribute to that same end. For Galen and Fabricius, these *utilitates* of parts are always ultimately dependent upon the *actions* of those parts. The actions and their contributions to the life of the animal are "of first importance," and so the *utilitates* of parts are articulated in terms of the ways the components and features of a part contribute to that action:

> For the *utilitates* of an organ always have reference to its action, and depend upon the action which proceeds from a homogeneous part of it. For this reason, in every organ there is always provided one part from which the action proceeds, while the other parts of the organ are related to the action as useful assistants.[24] (79)

In this way, the action serves as final cause of the various components and features of a part.[25] And Fabricius's anatomical works typically culminate in a final section on the *utilitates* of the part, in which he exhibits final causal explanations of the features and components of the parts (described in the first section, on *historia*) in terms of their contribution to that part's action (identified and traced to a principal part in the second section, on action).

The work on muscles is no exception. Fabricius begins by identifying the tripartite structure of his discussion of muscles:

> First then we will take our beginning from the fabric, *historia*, dissection, structure or composition of the muscle, … make clear its action, and disclose every *usus*.[26] (383)

[23] May, *On the usefulness*, 724.

[24] This translation is adapted from Adelmann, *Embryological Treatises*, 276.

[25] Fabricius says as much on numerous occasions—e.g., "The third part [of the treatise] being the part that pursues the *utilitates* … which always look toward and contemplate the action of the larynx (that is, voice) and *are directed to that action as towards an end*" (79; emphasis added).

[26] This is a rare instance where Fabricius uses *usus* instead of *utilitas* for the Galenic concept. (Fabricius was somewhat unusual in his preference for *utilitas*; most authors of the period preferred *usus*.)

In the section on the *historia* of the muscles, he discusses the subparts of the muscles, like flesh, membranes, arteries, veins, nerves, and tendons, as well as the constitution of muscle, especially out of flesh and *nervaceum corpus*. He also discusses variation in these parts as they constitute different muscles. He finally turns to features belonging to muscles as whole instruments:

> Because the fabric of the muscle is constructed from similar [i.e., homeomerous] parts, the muscle is constituted an organ[27] which requires a certain magnitude, shape, position, number, connection, insertion, and other differentia of an organ and dissimilar body, the treatment of which is presently laid out. (393)

Here Fabricius discusses the important variations exhibited by muscles in shape, dimensions, number, and in the placement of the origin and end of the muscles, as well as of the tendons. This leads eventually to a discussion of the presence or absence of muscles (or something analogous) and their major variations in all animals, divided into four categories or grades.

This preoccupation with patterns of presence, absence, and variation in muscles across all animals in his *historia* of muscle is also characteristic of Fabricius's approach to anatomy and reflects the Aristotelian aspects of his project. For Fabricius, the scope of anatomy unambiguously includes all animals. Although Galen discusses animal anatomy occasionally in, for example, *De usu partium* and makes comparisons between humans and other animals, and thinks the same distinctions between part, action, and usefulness apply to animals, his focus is still firmly on the human case and is medical in character. But when we turn to Fabricius, we see animal anatomy appearing for its own sake and with many of the comparative preoccupations found in Aristotle's animal books, texts that Fabricius frequently references.[28] James Lennox and Allan Gotthelf have done much to illuminate the methodological significance of the comparative aspects of Aristotle's zoology, connecting them with Aristotle's theory of scientific explanation in *Posterior Analytics*.[29] There, Aristotle

[27] An organ, unlike a similar part, has a particular shape, size, structure, position, etc. proper to it. The distinction is between, e.g., the liver and blood. To be a liver, a part must have a particular shape, etc. In contrast, blood does not require any particular shape or size.

[28] Indeed, Fabricius suggested that his complete body of work should be entitled *Totius animalis fabricae theatrum* (Fabricius *De visione*, dedication to *De aure auditus organo liber*). I provide a more detailed discussion of Fabricius's project than is possible here in "Descartes' 'Teleo-mechanics in Medical Context." Gianna Pomata briefly discusses Fabricius's understanding of *historia* and helpfully locates it in the context of different and changing understandings of *historia* in 16th-century Europe; see Pomata, "*Praxis Historialis*: The Uses of Historia in Early Modern Medicine," in *Historia: empiricism and erudition in early modern Europe*, eds. Gianna Pomata and Nancy Siraisi (Cambridge, MA: MIT Press, 2005), 105–146.

[29] James Lennox, *Aristotle's Philosophy of Biology* (Cambridge, UK: Cambridge University Press, 2001), esp. chapters 1 and 2; Allan Gotthelf, *Teleology, First Principles, and Scientific Method in Aristotle's Biology* (Oxford, UK: Oxford University Press, 2012), esp. chapters 9 and 12.

insists that proper scientific explanations must occur at what Lennox calls "the correct level of generality"; in such explanations, the subject, attribute, and cause will be appropriately related both extensively and intensively. Lennox and Gotthelf have shown in some detail how this requirement shapes both Aristotle's explanations in *Parts of Animals* and his reports in *History of Animals*. Fabricius embraces this aspect of Aristotle's project, and it similarly shapes his published works. When studying a particular part, Fabricius tries to carefully delineate which animals do and do not have the part and how the part varies from species to species. Furthermore, Fabricius follows Aristotle in articulating these correlations in terms of shared and varying attributes (e.g., four-legged, blooded, or respiring). With these correlations in place, Fabricius seeks teleological explanations for the presence, absence, and systematic variation of the part under consideration. These explanations make reference to the action of that part, the usefulness of that action, and variations in the life and other features of the animals (i.e., by reference to those shared and varying features, in terms of which the patterns of presence and variation were articulated).

Before Fabricius turns to exhibit such teleological explanations of the presence and variation of muscles across animal kinds, he first identifies and discusses the action of the muscle. Besides the correlations articulated in the *historia* section, the action of the part features centrally in these explanations. Fabricius identifies the action of muscle as contraction or tension, and he identifies what he calls the "*nervaceum corpus*" (i.e., the fibers, not the flesh) as the principal part responsible for this action. He also discusses how this contraction contributes to the life of the animal (i.e., he discusses the *utilitates* of the *action*). Here he stresses the wide variety of activities to which muscle contributes by its contraction, including, for example, progression, eating, looking, and holding and expelling waste. Ultimately, suggests Fabricius, all of these activities are, in one way or another, at the service of pursuing what is beneficial and avoiding what is harmful.

This variety of ways in which individual muscles contribute to the life of the animal by their contractions grounds the teleological explanations of the variety found in muscles in the final section, *De musculi utilitatibus*. Here Fabricius discusses the subparts and features of the muscle, following the same general order of exposition he used in the section on the *historia* of the muscle. He begins with the way the various subparts of muscles (i.e., veins, arteries, flesh, nerves, and tendons) render them fitted to its identified action—that is, he traces the *utilitates* of the components of muscles. He then turns to consider the *utilitates* of muscles as wholes:

> Having explained the *utilitates* of the parts of muscle, it remains to explain the *utilitates* of those features that belong to the whole muscle. Now position, magnitude, connection, figure, number, and insertion

> do so belong; from all of these as from proper sources *utilitates* are acquired. (414)

It is here that Fabricius introduces mechanical principles. What role do these principles play in the discussion of the *utilitates* of muscles? These principles, I suggest, are primarily used to articulate the hypothetical necessity according to which certain features of the muscles are required, given the role the muscle's action plays in the life of the animal.[30]

In order to see this, and to see how it is simply one instance of a general approach in which material considerations are invoked to articulate the hypothetical necessity of features of parts, I first look at his discussion of the *utilitates* of the veins in muscles. I then compare this to his discussion of *utilitates* of the insertion points of muscles, where mechanical principles are invoked. The discussion of the veins is brief and can be quoted in full:

> Now it is said that the veins and arteries (especially the veins) come together in the fabric of the muscle so that the muscles, like a plant, may be nourished and continue living. Now this is especially so for muscle, as much as it receives and demands much larger veins and arteries than the other parts of the body. For where a greater portion of fleshy, hot, loose and soft substance is lost, it was appropriate that abundant blood be available and large veins lead for the purpose of restoring [that portion]. But if one of the parts of muscle is observed to be fleshier, or another dedicated to more frequent motion, then for this same reason also more frequent and larger veins will be provided to them. (409)

The veins, in general, distribute nourishment to the parts of the animal. In explaining their presence as components of muscle, Fabricius identifies the way this renders the muscle fitted to its action. The veins provide the necessary nutriment for keeping the muscle alive. That the muscle receives in general larger and more veins than other organs is explained with reference to muscles being especially prone to losing a "greater portion of fleshy, hot, loose and soft substance," and so in greater need of nourishment. The muscles tend to lose such material for two reasons, it seems. First, they are often very fleshy; second, when contracting they tend to lose material. This also explains the variety in the size and number of veins leading to different muscles. The more flesh in the muscle and

[30] For an account of Aristotle on "hypothetical necessity," see John Cooper, "Hypothetical necessity and natural teleology," in *Philosophical Issues in Aristotle's Biology*, eds. Allan Gotthelf and James Lennox (Cambridge, UK: Cambridge University Press, 1987), 243–274.

the more frequently it is used (the more frequently it performs its action), the larger and more numerous the veins.

How frequently a muscle will contract is connected, in turn, to what contribution it makes to the life of the animal. As his discussion of the *utilitates* of flesh in muscle makes clear, for Fabricius, flesh contributes in various ways to the strength and facility with which a muscle contracts and to the protection of the muscle's principal part. Muscles vary in their need for this strength, facility, and protection depending on the role they play in the life of the animal. So, like the variation in the size and number of veins leading to muscles, the variation in fleshiness of muscles will be explained with reference to the activities to which the muscles' contractions contribute. In both cases, the role the muscle plays explains and renders necessary certain features of the muscle. This hypothetical necessity is grounded in material considerations like the behavior of fleshy, hot, loose, and soft material in the circumstances of contraction. Given these features of the material and behavior involved, a certain fleshiness and provision of veins is rendered necessary, given the role of a particular muscle.

This same structure will be present in the consideration of the *utilitates* of the features of the muscles as wholes. One of these features is the insertion point of the muscle in the bone it moves. Fabricius says muscles have varied insertion points:

> For some are inserted into the beginning of the moving bone, others beyond, that is, either up to the middle, nearly into the middle, or beyond the middle. For the muscles which bend the second and third joints of the digits, likewise those of the elbow and of many others are inserted into the beginning of the bone to be moved. The deltoid muscle is inserted into the middle; the muscle drawing the arm to chest into short of the middle; the internal muscle of the femur, beyond the middle of the moving bone. (416)

After characterizing the variation under consideration, he goes on to identify how these variations fit the various muscles to their particular actions.

> The *utilitas* of these variations is as follows: whenever nature sets up simply to move a bone, she always makes the insertion and the end of the muscles towards the beginning of the moving bone, as is the case in most muscles. Whereas if beyond the motion of the bone to be moved, nature also requires a strong motion, then she establishes the insertion beyond the beginning of the bone to be moved: and the further from the beginning the insertion is, the stronger the motion following. (416)

The variation in needed strength of the motion of the moved bone explains the variation in the insertion point. Fabricius traces this need for strength in the motion, in turn, to the role that motion plays in the life of the animal:

> At this point it is appropriate finally to understand that [among] strong motions, some are going to be strong from themselves and through the proper nature of the action of the member (as are the first and second muscles flexing the digits of the hand); others are going to be strong for another reason, evidently on account of the weight of the bone or the member to be moved (for example, such are the muscles of the femur, the humerus, and the shoulders); finally others are going to be strong because of each cause. For example, such are the muscles of the elbow and knee; for by the elbow large and heavy bodies are lifted and raised; in addition, the remainder of the arm is raised and carried—and similarly with the knee. (416)

The strength of the motion is required because the muscle moves a larger part of the body in addition to the bone it moves directly, or because the bodily motion itself must be strong, or for both reasons. Thus, the role the action of the muscle plays in the animal explains and renders the strength of the motion produced by the muscle necessary or best. This needed strength of motion, in turn, explains and renders necessary the insertion point. In order to understand the necessity of the final step—in order to grasp the reason why the further out the insertion point, the stronger the motion—one must bring in material considerations. In this case, as we shall see, these considerations include mechanical principles.

The insertion point is one of three features that appear in the body of his discussion briefly, but to which Fabricius gives special attention at the end of the treatise. He opens this final discussion by listing the problems and stating that each has a mathematical and natural cause:

> In the last place there remain three problems, which have been mentioned briefly several times before, but which are to be discussed more exactly in this last part.
>
> 1. Why the origin of muscles is always more raised than its end.
> 2. Why muscles whose ends are not in the beginning of the moved bone, but further up move the moved bone more easily.
> 3. Why the longer muscles produce not only longer but also stronger motions.
>
> . . .
>
> A natural and a mathematical, or better mechanical, cause can be given for all of these. (416–417)[31]

[31] Fabricius adds a fourth problem to these three: "To these a fourth from Galen is added: why weights which we cannot move by hand, we are accustomed to set in motion and lift by applying levers or drag and transfer by an attached rope."

He then provides two discussions. In the first, he provides "natural" causes, and in the second, "mechanical" ones.

The "natural" explanation he gives for question 2 is the same as that provided earlier in the text when he first examines variations in insertion point. There he explained the correlation between further insertion points and stronger motions:

> The cause of this is that all the insertions of this kind inasmuch as they subtract part of the length of the bone and member to be moved, so the weight is subtracted and a shorter remaining member to be moved is delivered, and hence it is moved more lightly and easily. (416)

The claim here, then, is that the muscles with insertion points farther from the joint have, effectively, a smaller load to move. The idea must be that the muscle is only moving the part of bone that extends beyond the insertion point. It is this consideration that reveals the connection between stronger motions of limbs and insertion points, and so grounds the hypothetical necessity by which a certain role for the muscle determines the insertion point.

Fabricius returns to this question in his "mathematical" or "mechanical" discussion. Here too the causes identified are meant to ground the hypothetical necessity connecting the role of the muscle and its insertion point. However, the necessary connection is revealed by noting that the muscle and bone have the *ratio* of a lever. Fabricius suggests they have the *ratio* of a lever in at least two ways. He discusses each of these ways and also provides a third discussion of the limb as a lever, grounding the law of the lever in the nature of circular motion. In each case, in order to show how the limb has the *ratio* of a lever, Fabricius identifies the fulcrum, the applied force (*vis movens* or *vis deprimens*), and the weight to be moved or elevated (*pondus elevandum* or *pondus attolendum*). Fabricius treats the limb as an *inverted* lever, with the muscle insertion providing the fulcrum:

> Can it be that the solution to this question is to be sought from the nature of an inverted lever? Can it be then that the cause of this thing is that the moved bone is made just as a lever whose fulcrum is above the end of the muscle? Indeed, it can be considered a lever in two ways. (419)

The first way in which the moved bone can be considered a lever locates the *vis deprimens* in the beginning of the moved bone (the end terminating in the joint flexing). He identifies it as the weight of the bone or at least of its extremity (*gravitas ipsius osis movendi, vel etiam gravitas ossis extremi*). The weight to be elevated he identifies as the whole load of the bone from the insertion point to the end of the bone (*tutum onus ossis movendi, a fine musculi ad extremum usque*). He then says that the farther the insertion point is from the joint, the longer and therefore heavier is the *vis deprimens* and the shorter and therefore lighter is the *pondus elevandum*. Thus, the farther the insertion point, the easier it is for the beginning of the bone to elevate the rest of the bone (419). Although this account is similar to the natural explanation in treating the portion of the bone extended beyond the insertion as the load, it differs from that natural explanation in treating the insertion point as the fulcrum and the weight of the bone from the joint to the insertion point as the moving force. In the natural explanation, there was no lever; there was simply the muscle moving a load.

In describing the second way the moved bone shares the *ratio* of a lever, Fabricius exchanges the moving force and the weight to be elevated. Now the *vis movens* is the heaviness of the end of the moved bone and is in that end (*vis movens sit in fine ossis movendi, ipsius videlicet gravitas*). The beginning of the bone and the adjacent parts are then the weight to be lifted (*pondus attolendum*). He then says that in a lever, the moving force is more powerful the farther it acts from the fulcrum. This, he says, means that the (effective) heaviness of the end of the bone will be greater the farther it is from the insertion point. And this, he claims, means that that end of the bone will press the fulcrum more heavily (i.e., it will press down on the muscle more heavily). Conversely, the closer the insertion point is to the end of the bone (in other words, the farther it is from the joint), the lighter will be the weight of the bone on the insertion point. "So on that account, the further beyond the beginning of the moved bone the end of the muscle is, the easier will the moved bone be moved" (419). It is worth noting that, unlike the first way of considering the moved bone as a lever, here mechanical advantage is involved. In the first case, no change in mechanical advantage was identified, only a change in the magnitudes of the moving force and the weight to be elevated.

After providing these two treatments of the bone as a lever, he rehearses briefly the Aristotelian grounding of the law of the lever, and identifies the insertion point with the center of circular motion and so the fulcrum:

> Furthermore, all these things are sought from the nature of the circle. Certainly the longer a line is from the center, the faster it is carried, and thus more easily raises the shorter line produced beyond the center [i.e., on the other side of the center]. Therefore let the end of the muscle be the center, the parts of the bone extending on both sides be the lines proceeding from the center, the longer of which will more easily raise the other. (419)

He then goes on to claim that his analysis of the bone as lever is acceptable even if the fulcrum is not fixed but is moved while one end of the lever remains still:

> Nor is it important to this analysis whether one part of the lever is pressed down and the other raised by means of the fixed still fulcrum of the lever, or whether the fulcrum moves itself and one part of the lever by means of the fixed still beginning of the other part of the lever, being endowed with a force for the purpose of raising itself. (419)

He is suggesting that the two situations are equivalent. These he presents using two diagrams (see figure 4.1). In the upper diagram, we have a standard lever, with fulcrum L. The applied weight H presses at F, elevating G. This is then said to be equivalent to the situation in the lower diagram. Here we have, says Fabricius, an inverted lever with the fulcrum at M. The portion MB is to be elevated by some moving force. However, this is accomplished not by a force being applied on the other end, but rather by having that end fixed (by contact with D); the motion is accomplished by the fulcrum raising itself from M to E, elevating MB and incidentally also the part from M to D. This latter instance, presumably, is supposed to be the situation with a joint. The fulcrum is the insertion point of the muscle into the moved bone, which makes a joint with another

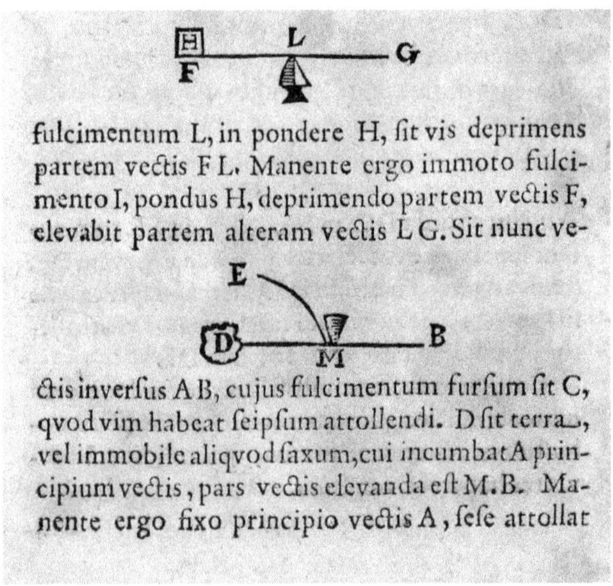

fulcimentum L, in pondere H, ſit vis deprimens partem vectis F L. Manente ergo immoto fulcimento I, pondus H, deprimendo partem vectis F, elevabit partem alteram vectis L G. Sit nunc ve-

ctis inverſus A B, cujus fulcimentum furſum ſit C, qvod vim habeat ſeipſum attollendi. D ſit terra, vel immobile aliqvod ſaxum, cui incumbat A principium vectis, pars vectis elevanda eſt M.B. Manente ergo fixo principio vectis A, ſeſe attollat

Figure 4.1 Hieronymus Fabricius ab Aquapendente, *Opera omnia anatomica & physiologica* (Lipsiae, DE: Sumptibus Johannis Friderici Gleditschii, Excudebat Christianus Goezius, 1687), 419. Credit: Université de Strasbourg, Service Commun de la Documentation (France).

bone at D. The muscle contracts and, by means of its attachment to D, elevates the insertion point and the whole bone.

At this point it is not important to consider the skill with which Fabricius attempts to analyze the movement of the limb in terms of levers, or evaluate his equivalence claim.[32] What is important for my purposes is to return to the question of the role of mechanics in Fabricius's work. The mechanical principles are invoked in order to reveal the cause of the connection between insertion point location and ease of motion. They are grounding the hypothetical necessity connecting the function of a muscle and several of its features. Thus, the mechanical principles play the same kind of role as the principles of matter theory grounding explanations of the usefulness of hot, moist flesh surrounding the contracting fibers of the *nervaceum corpus* or of the presence of large veins in fleshy muscles. Mechanical principles, like those of matter theory more generally, are being integrated into the teleological explanations of part variations at the heart of Fabricius's project in *De musculi utilitatibus,* and indeed in all his discussions of the *utilitates* of parts.

It should be noted that since these mechanical principles are grounding the correlation between the insertion point and the strength of motion, they *could* also be invoked in explanations of the strength of that motion. Given the insertion point and other relevant anatomical facts, one could use mechanical principles to provide such an explanation. Although this *could* be done, this is not what Fabricius *actually* does. Scholars like Jaynes and Baldini, by failing to appreciate Fabricius's Galeno-Aristotelian anatomical project, misinterpret Fabricius's use of mechanics, perhaps tacitly assimilating his efforts to a story of a 17th-century rise of mechanism and the decline of Aristotelianism and its teleological modes of explanation.

4. Fabricius, Aristotelianism, and Mechanics as a Subordinate Science

But for Fabricius, there is no essential tension between the application of mechanics and the employment of Aristotelian resources. Indeed, Fabricius seems to take his mechanics to be a *part* of his Aristotelianism and to be rooted in the *Quaestiones Mechanicae,* which Fabricius attributes to Aristotle. Fabricius explicitly references this text at numerous places and employs its grounding of the law of the lever in the properties of circular motion previously

[32] The equivalence claim fails, and the whole discussion seems crippled by his identification of the insertion point as the fulcrum.

mentioned.[33] It is worthwhile, then, to look briefly at the Aristotelian understanding of subordinate sciences in general and mechanics in particular. This will shed light on how Fabricius could see the application of mathematical mechanics to anatomy as an extension of his Aristotleianism.

The *Quaestiones mechanicae* is constituted by a preface, which discusses the scope and character of mechanics, and 35 separate questions or problems.[34] The problems vary from questions regarding the balance, lever, and pulley to treatments of features of ships, forceps, nutcrackers, and projectiles. Interestingly, it also includes a discussion of how we stand up. The preface delimits mechanics as the art that aids us in overcoming the difficulties associated with accomplishing tasks for our benefit that are contrary to or beyond [*praeter*] nature, particularly by moving greater weights with smaller forces. It also says that mechanics shares something with both natural and mathematical theorizing. The "how" of mechanics is made clear by mathematics, though the "about" or "with respect to what" is made clear by natural philosophy. This seems to mean that the actual objects and devices considered by mechanics are natural (i.e., made of matter and having natural motions, etc.), but that the causes of the motions involved in achieving our tasks are made clear by mathematical considerations.

This characterization of mechanics as sharing in both natural and mathematical science marks it as an Aristotelian subordinate science and connects it with a number of passages in the Aristotelian corpus that treat these sciences (astronomy, harmonics, optics, and mechanics are the examples found in these passages).[35] In the *Metaphysics*, Aristotle says that these sciences are more precise than others because they consider only certain features of their objects:

> The same account may be given of harmonics and optics; for neither considers its objects *qua* sight [ὄψις] or *qua* voice, but *qua* lines and numbers; but the latter are attributes proper to the former. And mechanics too proceeds in the same way. (1078a14–17)[36]

[33] See, e.g., "An ergo de vecte bene respondet Arist. q. 3. Mechanica?" (419); and "...juxta Aristotelem in Q. 29. Mechanica." (420). For the grounding of the law of the lever, see 419. Fabricius also refers to this approach to grounding the law of the lever in his related work on joints; see 429.

[34] For an orientation to the reception of the *Mechanicae* in the 16th century, see Paul Laurence Rose and Stillman Drake, "The Pseudo-Aristotelian *Questions of Mechanics* in Renaissance Culture," *Studies in the Renaissance* 18 (1971): 65–104; and W. Roy Laird, "The Scope of Renaissance Mechanics," *Osiris* 2 (1986): 43–68.

[35] I provide a more detailed discussion of Aristotle's view of the subordinate sciences in "The Intersection of Mathematical and Natural Science: The Subordinate Sciences in Aristotle," *Apeiron* 46.2 (2013): 85–105.

[36] Translation adapted. Mechanics is also explicitly mentioned in the treatment of the subordinate sciences found in *Posterior Analytics*; see 78b35f.

These sciences, although they are about natural objects like voice and sight, do not consider sight and voice qua natural (qua sight and voice). Such a consideration would require a treatment of the faculties of vision and hearing, the causes of sight and voice, their medium, etc. Optics and harmonics treat their subjects only inasmuch as they exhibit mathematical attributes, that is, qua line and number. The features of voice and sight that are explained by harmonics and optics are such that their causes are made clear by mathematics. That mechanics "proceeds in the same way" is reflected in the *Quaestiones Mechanicae*.

> The original cause of all such phenomena [of the lever] is the circle. It is quite natural that this should be so; for there is nothing strange in a lesser marvel being caused by a greater marvel, and it is a very great marvel that contraries should be present together, and the circle is made up of contraries. (847b15–19)

It is also explained that

> [t]he phenomena observed in the balance can be referred to the circle, and those observed in the lever to the balance; while practically all the other phenomena of mechanical motion are connected with the lever. (848a11–14)

Thus, the explanation of almost all mechanical phenomena will be given in terms of the circle (or circular motion) and its properties (i.e., in mathematical terms). We have already noted that Fabricius refers the behavior of levers to the nature of the circle, embedding himself in this tradition.[37]

It is worth emphasizing that for Aristotle, sciences like mathematical optics and mechanics are understood to have a limited but genuine scope—an Aristotelian approach to the mathematical sciences is realist but not reductionist or eliminativist. The natural objects being considered do not exhibit only mathematical properties, but they *do* exhibit those mathematical properties considered by the

[37] Fabricius's own longtime colleague at Padua was mathematician Guiseppe Moletti. In his study of Renaissance treatments of the *Quaestiones Mechanicae*, W. Roy Laird examines Moletti's lecture notes. He characterizes Moletti's understanding as follows: "The subject of mechanics, Moletti concludes, is not simply machines, but rather sensible quantity mobile in circular motion or compounded from it. The properties of mechanics are the powers and virtues of such machines for lifting and drawing weights and for throwing projectiles" (Laird, *Scope*, 61). For more on Moletti and his suggestive attempts to connect mechanics and animal locomotion, see also Laird, "Nature, Mechanics, and Voluntary Movement in Giuseppe Moletti's Lectures on the Pseudo-Aristotelian *Mechanica*," in *Mechanics and Natural Philosophy Before the Scientific Revolution*, eds. W. Roy Laird and S. Roux (Dordrecht, DE: Springer, 2008), 173–184.

science. This aspect is reflected in Fabricius's approach. Throughout his work, Fabricius rallies much of Galenic and Aristotelian natural philosophy, invoking the elements, elemental qualities, temperaments, faculties, souls, etc. There is no evidence that Fabricius thinks the employment of mathematical mechanics is aimed at replacing explanations of phenomena employing these more traditional natural philosophical resources. Nor does he seem to think he will provide more fundamental accounts of these resources in mechanical terms. In addition, he does not see any tension or dichotomy between the employment of mechanical principles and the use of teleological explanations.

Fabricius seems to think, instead, that mechanical reasoning provides genuine knowledge of certain necessary connections between structures and the effect of moving forces applied to those structures. These necessary connections will be relevant to animal anatomy because certain parts of animals (muscles and bones) exhibit these structures and are, furthermore, the soul's principle and proximate instruments for causing motions by the application of moving forces. For these reasons, mathematical mechanics will be employed in teleological explanations of those parts, revealing the hypothetical necessity, according to which the specific functions the instrumental parts play entail and explain the presence and features of those parts.

5

Metaphysical Problems in Francis Glisson's Theory of Irritability

ANNE-LISE REY

1. Introduction

Francis Glisson is well known for having highlighted, both in his *Tractatus de Anatomia hepatis*[1] (1654) and especially later in his *Tractatus de ventriculo anatomis* (1677),[2] the notion of the *irritability* (*irritabilitas*) of the fiber, understood as an internal force of action and reaction.[3] Specifically, whereas in the text of 1654, Glisson still considers that irritation depends on "putting the nerves on alert" (*vigoratio*), in 1677, he characterizes the notion of irritation without making it depend on the movement of the nerves or the awareness of the human being. Thus, it becomes a generic concept to account for an internal force at work in the body, which supposes a perception without awareness and is also a kind of perception before sensation.

Between these two texts, whose comparison helps us understand the evolution of his conception of irritation—that is, from irritation to irritability—Glisson wrote his theoretical treatise *De natura substantiae energetica* in 1672

[1] In the *Anatomia hepatis* (Londini, UK: Du-Gardianis, 1654), Francis Glisson uses the word "irritatio" several times, particularly in chapter XLIV, 396, and chapter XLV, 454, and also "irritabilitas" (481).

[2] In the *De ventriculo et intestinis* (Londini, UK: E. F., 1677), Glisson this time uses the word "irritabilitas," particularly in chapter IX, "De pauculis irritabilitatis differentiis," §7, 197–199, and to explain the different levels of irritability in chapter XXIV, "De facultate excretrice, sive expultrice," §21, 472.

[3] More precisely, in *De ventriculo*, chapter V, "De Fibris in genere," §15 and 16 (142), Glisson distinguishes between "actio fibrae" and "passio fibrae," and he identifies them respectively to "motus activus" and "motus passivus." Cf. on this question Jean Starobinski, "La réaction et la machine animale: Hobbes, Glisson, Buffon," in *Action et réaction: vie et aventures d'un couple* (Paris: Editions du Seuil, 1999), 205–220.

and also reworked his address to the reader for the treatise of 1677, which he had started several years earlier. It should be remembered, as demonstrated by Guido Giglioni,[4] that in those years, Glisson was guided by the need to make concepts match things and to resolve, as a necessary preliminary measure, the "big question" of the general perception of Nature.

The great impact this conceptualization had on the work of Haller[5] and on the Montpellier doctors in the 18th century is well known; less well understood is, first, what led Glisson to extend the scope of irritability, and second, whether and if so, how his philosophical and epistemological positions contributed to this extension.

I would like to highlight two points and ask how they are connected. First, it seems to me that Glisson's model of the body can be considered as something like an empirical model. I mean by 'model' the means of giving more intelligibility to reality by using a fruitful analogy. In this case, by 'empirical model,' I mean a way of explaining the actions of the body by an empirical conception of cognition. In his article "Il 'Tractatus' de Francis Glisson,"[6] Giglioni has shown that in the treatise of 1672, any kind of mechanistic explanation is explicitly rejected, and that this rejection can be understood as an indication of the influence of Bacon on Glisson. It seems to me that Giglioni is talking about the cognitive method applied by Glisson, namely, the denial of any external causality to explain the production of the first principle of life, that is, the internal principle of perception, appetites, and movement, in §10 of the address to the reader.[7] Correspondingly, the "requirement" to apply an epistemological principle inspired by Bacon led him to constantly resort to experience, as §6 shows.[8]

[4] I have in mind the great text of Glisson edited by Guido Giglioni, "Tractatus de Inadaequatis Rerum conceptibus" (Sloane 3314, ff. 8r–168r) in *Latin Manuscripts of Francis Glisson (1) Philosophical Papers. Material related to De Natura Substantiae Energetica (On the Energetic Nature of Substance)*, 1672 (Cambridge, UK: Wellcome Texts and Documents, 1996), 5–114.

[5] As indicated by Giglioni in his illuminating paper "What ever happened to Francis Glisson? Albrecht Haller and the Fate of Eighteenth Century Irritability," *Science in Context*, 21, issue 4 (2008), esp. 466, quoting Haller, 1757–1766; *Elementa physiologiae corporis humani*, 8 vols. (Bern and Lausanne, CH: Bousquet, D'Arnay, Grasset, IV, 461): "Franciscus Glisson, qui universis elementis corporum vim motricem tribuit, etiam nostram vim Irritabilitatem vocavit, non quod absque irritatione nunquam adpareat, sed quod ab irritatione certa succedat."

[6] G. Giglioni, "Il Tractatus de Natura substantiae energetica di F. Glisson," *Annali della Facoltà di lettere e filosofia dell'Università di Macerata*, 24 (1991): 137–179, 178.

[7] F. Glisson, *Tractatus de Natura substantiae energetica, seu de vita naturae* (1672), Ad Lectorem §10: "...Est enim, internum principium percipiendi, appetendi, atque se movendi, ei indere. Cujus productionem nulla vis externa, nullus motus aut quies, nulla textura, schematismus, organisatio, proportio, nexusve partium, attingere queat."

[8] F. Glisson, *Tractatus de Natura substantiae energetica, seu de vita naturae* (1672), Ad Lectorem §6: "...Eliciuntur autem testimonio sensuum, nempe, sensatione circa eas multis modis, ne subsit fallacia, etiam adhibitis adminiculis, variata; experientia multiplici; observatione sedula, analogismo (cui soli rarius fidendum; imo nunquam, nisi ubi aliis mediis destituti fuerimus) argumentatione, cui

My hypothesis, however, is slightly different: I am interested in analyzing how Glisson explains the activity of the organs by using cognitive modalities. In this context, Walter Pagel[9] has stressed the influence of Van Helmont on Glisson. Starting from Van Helmont's notion of 'Archée,' Glisson explains the action of nerves along the lines of the model of the operations of the mind. To say that the organ is, in a sense, "like a mind," seems to imply two questions: (1) What kind of mind? Or, more precisely, which conception of the mind? And (2) Why should this conception of the organ, whose functioning can be made intelligible by analogy with the functioning of the mind, not also be understood to some extent in terms of the model of the body? The latter question may seem at best paradoxical and at worst a tautology. But does this first attempt at relocating the active principle so that it is no longer, or at least not solely, in the brain but also, to some degree, in the nerves, not establish a theoretical space in which the body can be conceived "as a body," though indeed no longer as a body reduced to mechanical operations? Here, it is rather conceived as a body with an active principle of its own, which therefore requires some form of autonomy with respect to what Glisson calls the "big order" brain.[10]

This might provide a possible explanation for the function of natural perception in the motor activity of the body. It might supplement the hypothesis of a direct connection between Glisson and the work of William Harvey, and it might therefore help to establish the influence of Harvey's conception of the vegetative soul on Glisson. I would further hypothesize that the way the issue of animal sensibility was resolved in the late Middle Ages, through the development of the idea of a sensitive cogitative faculty, could perhaps yield some very fruitful insights into this question.

In fact, my central hypothesis is that the requirements established by the theory of knowledge that Glisson was to deploy in his various texts, as well as his conception of perception, led him to adopt a wider notion of irritability. Thus, the philosophical requirement for coherence in his natural philosophy led him to change the definition and scope of the concept of irritability.

caute quoque subscribendum est. Verum si certissima ac tutissima inveniendi veritatem methodo uti placuerit, consule Novum Organum Cl. D. Baconii, cujus doctrinae vestigiis si fideliter institeris, haud multum, aut facile, a vero aberraveris."

[9] W. Pagel, "Harvey and Glisson on irritability with a note on Van Helmont," *Bulletin of the History of Medicine*, XLI, 6 (1967): 497–514.

[10] F. Glisson, *Anatomia hepatis* (1654), chapter 44: "... Si vasa fellea sensu omni destituentur, non possent certe ad hunc modum irritari. Quippe irritatio omnis subesse perceptionem indicat: quicquid enim lacessitur, id illico excitatur, ut se ab illata injuria vindicet; quod sane, nisi percepta aliquatenus re laedente, fieri nequaquam posset. Ideoque, si aequum erat, ut vasa fellea essent irratibilia; nervorum quoque in illa distributione opus erat, a quibus sensum mutuarentur."

2. The Movement from Irritation to Irritability

The meaning of the concept of irritability in Glisson's thought has been analyzed by many scholars. Two points are particularly worth keeping in mind. As Owsei Temkin showed in his now-classic article of 1964 entitled "The Classical Roots of Glisson's Doctrine of Irritation,"[11] contrary to what Haller said, the first instance of irritation cannot be assigned to the text of the *Tractatus de ventriculo* of 1677, since it can be found in the *Anatomia hepatis* of 1654, and the theme of irritation is even addressed in the *De rachitide* of 1650.

Temkin also pointed out that the distinctive hallmark of Glisson's thought is not so much that he identified, in the specific context of a treatise on the anatomy of the liver, a property of organs explaining what drives the bile ducts to get rid of bile from the liver. Glisson, of course, identified irritation as the cause of the evacuation of secretions, specifying three possible causes of irritation: the fact that the ducts are full, that they contain an acrimonious or bitter bile, and, finally, that a certain quickening or setting off of the nerves (*vigoratio nervosum*) leads to this evacuation.

It seems to me, however, that Glisson's exploration of the precise relationship between perception and what incites organs to move is also significant, to the extent that he proposes a new meaning for the concept of perception. One could say that following his intellectual journey means discerning, through the writings of one author, the path leading from perception, understood as sensitive perception and defined as a property of nerve fibers, to a perception that is not necessarily sensitive but rather a property of all living beings: their ability to act and react.

We must therefore consider the status of irritation in the text of 1654 from the wider perspective of a study that follows the progressive development of an idea, aiming to understand with which theory of knowledge it is associated, and how it conceives its relation to the mechanistic model.

When the notion of irritation appears in chapter 44 of Glisson's *Anatomia hepatis*, it is deployed in different ways—irritable, stimulated, irritated, and irritating—and it is necessarily linked to perception ("in fact, each irritation indicates the existence of a perception").[12] I call this perception sensitive, since it is consistently linked to the presence of nerves,[13] which are conceived

[11] O. Temkin, "The classical roots of Glisson's Doctrine of Irritation," *Bulletin of the History of Medicine* 38 (July–Aug. 1964): 297–328.

[12] F. Glisson, *Anatomia hepatis* (1654), chapter 44, "... Si vasa fellea sens omni destituentur, non possent certe ad hunc modum irritaru. Quippe irritatio omnis subesse perceptionem indicat...."

[13] In the context of the *Anatomia hepatis*, the title of chapter XLIV indicates clearly this link: "de actione et usu radicum cavae, porique bilarii; atque etiam nervi hepatici." Particularly 397: "Quomodo enim res dicatur irritari, quae neque jam actu movetur, nec ad motum ullo pacto accigintur? Ideoque ut vas fellea essent irritabilia, oportuit eadem nervis praedita esse, quibus motum talem perficerent,

as mediators. Furthermore, it is related to the operations of the brain, which induce an awareness of this perception. Glisson thus identifies three kinds of irritation that can trigger action in the nerves, leading to movements in the body. The three types of irritation are sensation, appetite, and movement of the muscles. This tripartition corresponds, point by point, to the theory of Galen.[14]

In this context, Glisson mobilizes specific explanatory models that are not necessarily cognitive, but that enable us to identify to which specific regime of intelligibility the irritation discussed in the *Anatomia hepatis* belongs. It seems to me that the definition of irritation in the *Anatomia hepatis* is situated between two systems of explanation, both present in the text of 1654: First, Glisson sets out the idea of the self-regulation of the gallbladder by means of a mechanical process of contraction and expansion that regulates the flow. Along the same lines, he evokes, using a Galenic conception, the idea of a faculty of attraction at a distance[15] to explain the coming together of similar parts, identifying this as similar to the relationship between iron and magnet. And finally, he develops a model of action and reaction that conceives the stimulation of the fibers as an answer to external stimulation (here again we find the idea of the awakening of the nerves). Thus Glisson can be said to have established, on the one hand, a mechanical explanatory system (both internal and external), while, on the other hand, beginning with this text, he was also claiming that there is a vital activity that is actualized in the body's pursuit of a final purpose.

In chapter 18, §19 of the *De natura substantiae energetica*, Glisson defines 'life' in these words: "...the energetic nature of substance and the internal principle of movement and operations."[17] This definition is offered in answer to a problem formulated in §1 of chapter 16, in which Glisson begins with a definition

qualem sensatio eorum postulabat." And "Qui motus etsi semper non sit voluntarius, non poterit tamen sine nervorum ope expediri, cum a sensatione dependeat."

[14] O. Temkin showed, concerning the association of the sensation and the nerves, the difference between Glisson's conception (he understands this association as necessary) and Galen's conception, which is more ambiguous and changing from *The Usefulness and the parts of the body* to *The natural faculties*. Temkin, "The classical roots of Glisson's Doctrine of irritation," 311).

[15] Galen, *On the natural faculties* (Loeb Classical Library, London: William Heinemann, 1916), book III, §§13 and 14.

[16] F. Glisson, *Tractatus de Natura substantiae energetica, seu de vita naturae* (1672), chapter18, §19: "Ex dictis elucescit, vitam primaevam non esse accidens, non facultatem aut vires vitales, non motum aut actionem, non terminum motus, non formam physicam; imo nec esse substantiam, prout ea concrete sumitur: esse tamen substantiam, prout ea concrete sumitur: esse tamen substantiam, prout eadem abstracte et inadaequate accipitur; scilicet, esse entitatem per se subsistentem, sed consideratam in ordine ad operationes; hoc est, esse naturam substantiae energeticam,et internum motus seu operationis principium."

of substance as "that which exists by itself and is endowed with a certain vital nature."[17] He goes on to claim that this vital nature corresponds to the three primary faculties: perceptual, motor, and appetitive. While he underlines the general agreement around this definition as a characterization of individual substances, he brings up, at the same time, something that may cause controversy: the matter of whether material substances can be endowed with a vital nature. Here, he appears to reveal his ambition: "It is our responsibility to prove that matter is not only capable of vital nature, but alive in act, that is to say, endowed with perceptive, appetitive, locomotive vital faculties."[18] The key, then, is to understand how Glisson explains his view that matter could be alive in act.

In this connection, it seems to me that John Henry's idea of explaining the passage from irritation to irritability through an analysis of the courses in anatomy given by Glisson at the Royal College of Physicians of London after the publication of the *Anatomia hepatis* is extremely fruitful. Henry has shown that though in his early lessons, immediately following the publication of the treatise of 1654, Glisson forcefully stated the case for the cerebral location of the three faculties of perception, appetite, and movement, so that the brain is clearly shown to be the seat of perception and appetite, the question of whether it is also the seat of our ability to move would soon become less straightforward:

> The manner how this power or faculty is put in an execution, is soe obscure and harde to be conceived, that the minde reflecting upon it selfe, cannot clearly trace its owne footstepps. It is hard indeed to follow the track of the idaea's from the outward senses to the braine, but much more difficult to trace the edicts of the braine back againe to the outward organs of motion, yet this in nature must be done.[19]

Thus we can understand at a first level of explanation that, since it was impossible to account for the motor activity of the body in a way that was centered in the brain, Glisson was forced to look elsewhere for an explanation of movement in the body, thereby somewhat decentering the active power. This led him

[17] *Tractatus de Natura substantiae energetica*, chapter 16, Vita naturae substantialis asseritur a priore; et primo, a causis externis; ab Efficiente, Exemplari, et Finali. §1, 217: "Dico igitur omnes substantias proprie sic dictas, hoc est, per se sive suo marte subsistentes, esse natura quadam vitali, nempe tribus istis primis facultatibus, perceptiva, appetitiva et motiva, praeditas."

[18] Glisson, *Tractatus de Natura substantiae energetica*, §2, 217: "Quare hoc nobis incumbit probandum, materiam esse non tantum naturae vitalis capacem, sed et actu vivam, vitalibus nempe facultatibus, perceptiva, appetitiva et motiva, praeditam."

[19] Quoted in J. Henry, "Medicine and Pneumatology: Henry More, Richard Baxter, and Francis Glisson's Treatise on the Energetic Nature of Substance," *Medical History* 31 (1987): 18.

to assert that the brain simply gives a 'general command,' a general stimulation, causing the nerves to stimulate the muscles that produce the movement. He thus conceived the nerves on the model of the mind, that is, as able to learn through habit and experience, which explains how the nerves, once they receive a general signal, know what to do and can therefore transmit instructions to the muscles, thereby producing movement:

> Those nerves onely, which are propper to the action required [since he distinguishes between nerves that are sufficiently stimulated to produce action and those whose stimulation produces no movement] have learned by custome and long practice that they are to move upon such a signal given, and the other nerves have in like manner bene accustomed to sitt still upon the same signall . . . wherefore, I conclude, that the nerves learne to execute the dictates of the braine, upon the signall given, by costume and often experience.[20]

This passage is very interesting, and not only insofar as it points to the hesitations, indeed contradictions, in the writings of Glisson. John Henry, after all, points out that Glisson is only following the example of Harvey. In fact, Harvey, when discussing the seat of perception in chapter 57 of his *De animalium generatione* of 1651, claims both that it is dependent on the brain and, a few lines earlier, that movements are independent of the brain and of the other sensory organs:

> Now this *sensorium commune,* to which the species or impressions of all the external instruments of sensation are referred, is obviously the brain, which, along with its nerves and the external organs annexed, is held and esteemed to be the adequate instrument of sensation. And this brain is like a sensitive root to which a variety of fibres tend, one of which sees, another hears, a third touches, and a fourth and a fifth smell and taste. But as there are some actions and motions the government or direction of which is not dependent on the brain, and which are therefore called natural, so also is it to be concluded that there is a certain sense or form of touch which is not reffered to the common sensorium, nor in any way communicated to the brain, so that we do not perceive by this sense that we feel.[21]

[20] F. Glisson, British Library, Sloane, MS 3306, f.165–166.

[21] W. Harvey, *De generatione animalium,* 57, 1651 (English trans. Robert Willis, Sydenham Society, ed. 1847).

This passage from Harvey's book is particularly interesting in its use of a *cognitive model* to explain the action of nerves, that is, in basing the explanation on the pattern of operations of the mind. Pagel, in his 1967 article,[22] showed that this conception of the nerves being able to learn from experience was probably borrowed from Jean-Baptiste Van Helmont, who considers that within the body all the fibers are able to "learn" and "know" through their *archaeus*, that is, their natural perception. Indeed, Van Helmont's *Ortus Medicinae* offers us a whole range of different mental powers by virtue of which the archaeus of an organ in a body can act: the *knowledge* that the pylorus has of what should remain in the stomach or what it deems necessary to evacuate, for example, or the capacity of the stomach to *discern* what is harmful, thus leading to vomiting, to name only a few.

But although the idea of using the lexical field of psychological operations to reflect the activity of organs can be traced back to Van Helmont, I am not sure that he is really the central reference. And it still seems to me that the cognitive model itself is different from Van Helmont's. Though this remains to be pursued in more detail, I would argue that in contrast to Van Helmont's use of a model that could very roughly be described as rationalist—to the extent that the organ can come to know, judge, discern, and learn, and is thus "like a mind"—Glisson appears to be using a different model that, again very roughly, could be described as empiricist. For him, the nerves can certainly learn, but only if they have acquired "experience" through long practice; their knowledge presupposes "habit."

It seems that Glisson's contribution is to use the cognitive model while disconnecting it from the mind, that is to say, while separating the analogy between the action of the nerves and the operations of the human mind on the one hand, from, on the other, the idea that this analogy is based on the fact that this action of the nerves arises from a mental principle that is the archaeus.

In this particular context of explaining the action of the nerves, I believe that the natural philosophy of Glisson is revealed in a different light. Indeed, when his first model of irritation, in the *Treatise* of 1654, was challenged, the vocabulary he used showed the conceptual instability with which Glisson was struggling: the famous unsuitability of concepts for things, like a thorn in his bed of theory. But it also showed a way to break away from this problem of unsuitability: by a sort of relocation of the active principle. By beginning to challenge the central function of the brain (although the notion of "large command" remains),

[22] W. Pagel, "Harvey and Glisson on irritability with a note on Van Helmont," *Bulletin of the History of Medicine*, vol. XLI, no. 6 (Nov.–Dec. 1967): 497–514, esp. 508–509.

Glisson introduced the idea, based on a cognitive empiricist model, of an active power at work in matter itself.

3. How Is Natural Perception To Be Understood?

Although natural perception, which allows us to establish and understand the concept of irritability, is clearly identified as having its origins in Galen and Harvey, I would like to analyze the idea of using models of the mind and body to explain the activity of the organs, in order to shed light, from another angle, on the relative autonomy of the action of the organs. For this purpose, I would like to recall a well-known aspect of Thomistic thought, namely, the idea of a sensitive cogitative faculty. I would like to consider whether the pervasiveness of references to Suárez in the *Treatise* might not be understood as a return to the position of Suárez on this issue, and not just as part of the metaphysical framework for the understanding of actualization.

As is well known, Aquinas presents the idea of a sensitive cogitative faculty to account for the possibility of synthesizing the sum of sensory experiences:

> The faculty that philosophers call cogitative is at the juncture of the sensitive part and the intellectual part, where the sensitive part touches the intellectual. Indeed, at one and the same time it uses the sensitive part to consider particular forms and it uses the intellectual part to gather data. That is why it is found only in man.[23]

Here what Aquinas is setting out (with the help of some neo-Platonic ideas) is the view that our sensory nature, while being both sensitive and material, is in a sense involved in the activities and functions of reason:

> The cogitative power is that which is highest in the sensitive part of man, and, thus, sense in some way comes in contact with the intellective part so that it participates in something of that which is lowest in the intellective part, namely discursive reason.[24]

[23] Super Sent., lib. 3d. 23q. 2a. 2qc., 1 and 3: "Ad tertium dicendum, quod illa potentia quae a philosophis dicitur cogitativa, est in confinio sensitivae et intellectivae partis, ubi pars sensitiva intellectivam attingit. Habet enim aliquid a parte sensitiva, scilicet quod consideret formas particulares; et habet aliquid ab intellectiva, scilicet quod conferat; unde et in solis hominibus est. . . ."

[24] Aquinas, *De veritate*, q.14, a.I, ad.9: "Ad nonum dicendum, quod potentia cogitativa est id quod est altissimum in parte sensitiva, unde attingit quodammodo ad partem intellectivam ut aliquid participet eius quod est in intellectiva parte infimum, scilicet rationis discursum. . . ."

It is this union between soul and body that gives the cogitative power the means to participate in intellectual activity, distinguishing it from the merely animal estimative power:

> The estimative, through which the animal seizes data that are not received in the senses, like attraction and repulsion, is inherent in the sensitive soul (of man) to the extent that the latter is in some way connected with reason.[25]

He further develops the idea of a cogitative faculty that has a specific organ in the body:

> Nevertheless, the mind has contact with singulars by reason of something else in so far as it has continuity with the sensitive powers which have particulars for their object. This conjunction comes about in two ways. First, the movement of the sensitive part terminates in the mind, as happens in the movement that goes from things to the soul. Thus, the mind knows singulars through a certain kind of reflection, as when the mind, in knowing its object, which is some universal nature, returns to knowledge of its own act, then to the species which is the principle of its act, and, finally, to the phantasm from which it has abstracted the species. In this way, it attains to some knowledge about singulars.
>
> . . . This conjunction is found in the movement from the soul to things, which begins from the mind and moves forward to the sensitive part in the mind's control over the lower powers. Here, the mind has contact with singulars through the mediation of particular reason, a power of the sensitive part, which joins and divides individual intentional likenesses—a power also known as the cogitative power—and which has a definite bodily organ, a cell in the center of the head.[26]

[25] *De veritate*, q.25, a.2, co.: "Vis aestimativa, per quam animal apprehendit intentiones non acceptas per sensum, ut amicitiam vel inimicitiam, inest animae sensitivae secundum quod participat aliquid rationis."

[26] Aquinas, *De veritate*, q.10, a.5, reply: "Et sic mens singulare cognoscit per quamdam reflexionem, prout scilicet mens cognoscendo obiectum suum, quod est aliqua natura universalis, redit in cognitionem sui actus, et ulterius in speciem quae est sui actus principium, et ulterius in phantasma a quo species est abstracta; et sic aliquam cognitionem de singulari accipit. Alio modo secundum quod motus qui est ab anima ad res, incipit a mente, et procedit in partem sensitivam, prout mens regit inferiores vires. Et sic singularibus se immiscet mediante ratione particulari, quae est potentia quaedam sensitivae partis componens et dividens intentiones individuales quae alio nomine dicitur cogitativa, et habet determinatum organum in corpore, scilicet mediam cellulam capitis." Trans. James V. McGlynn (Chicago: Henry Regnery, 1953).

Thus there is in Aquinas's thought the idea that human sensibility is distinct from animal sensibility, and that for this reason it can work to some extent, through its proximity to the mind, like a mind, by carrying out the same operations as the mind.

What this rough sketch of the Thomistic position highlights is the idea of a cogitative faculty capable of judgment at the heart of human sensibility. It seems that by choosing to situate the origin of this thesis in the medieval debate rather than in reference to Van Helmont, we gain the advantage of being able to point out the clear difference between what François Xavier Putallaz, in his book *Le sens de la réflexion chez Thomas d'Aquin*,[27] calls 'sensory reflection,' on the one hand, and on the other, the universal reason of the intellect. This difference is important, as is the fact that for Aquinas, though reflection is incomplete in the senses, it is perfect in the intellect, just as though judgment is sketchy in the senses, it is perfect in the intellect. Within this theoretical framework, Glisson's notion of the action of the organs by analogy with the functioning of a mind seems to be easier to grasp, but with two differences. First, the question of hierarchy does not arise in the same terms, since Glisson is not saying that the senses function like the mind but not as well. Rather, he is creating a mode of operation that permits us to understand how the organ functions, using the model of how the mind functions as a support. In other words, Glisson deals with the sensitive dimension of these operations not by claiming both proximity to the intellect and relative imperfection compared to the intellect, but rather as a way of producing a specific explanation of the action of the organs that signals a beginning of autonomy with respect to the cerebral big order. Second, as a corollary, he considers the experience whereby the organ "learns" to exercise or move, not as a "defect" in its own nature, but rather as the recognition of this specificity.

Glisson's reference to Suárez in the *De natura substantiae energetica* could be understood if we try to look and follow Suárez's position on the sensitive cogitative faculty. In his *De Anima*, Suárez wrote:

> For the sensitive powers seem to be participations in the intellective power. And it is especially of this that Saint Thomas seems to speak, but in addition these things do not have closer efficiency, but there is a sort of causality that serves as a model; for the sensitive cognition is like a certain imitation of the intellective one, in the way that the activities of

[27] François Xavier Putallaz, *Le sens de la réflexion chez Thomas d'Aquin* (Paris: Vrin, 1991).

> the vegetative part, to the extent that they are produced as in art and in the intellect, can be said to be imitations of the intellect.[28]

In other words, Suárez considers the sensitive knowledge to be like an imitation of intellective knowledge. This point seems to me important because it confirms the medieval origin of this question, thus situating Glisson's position in a specific intellectual context. To briefly summarize: The question of the capacity of the organ to judge is elaborated on the basis of the theory of vision and in other terms on the basis of the organ. The point is, of course, to establish that vision is an intellective activity to the extent that it always presupposes comparison, judgment, etc. The problem is to explain vision in general and to distinguish between animal and human vision. It is in this context that the theory of a sensitive cogitative faculty is elaborated. The point of the discussion is to determine the nature of visual illusions. Paradoxically, for some authors (e.g., Oresme),[29] the primitive sensation, vision, is not the origin of the error, but rather judgment. He thus concludes that the error is not in sense at all, but in the activity of the intellect.

But this point is also important because, thanks to this reference, it is possible to understand the specificity of Glisson's thought: He adopts the medieval framework and from this he develops something new. The central hypothesis of the *Treatise*, on the inadequation between concepts and things, which in the end Glisson decided not to publish, is to consider reflection on knowledge to be a sort of anatomizing. It is because the intellect dissects the objects of perception that it can understand them. It seems that, independently of the metaphorical level of an understanding of world as the result of dissection—in other words, independently of the understanding of knowledge as being like the practice of anatomy—two important points arise here. First, there is a willingness to conceive the spirit and the body within the same system of intelligibility. This implies a denial of pan-psychism, and it also implies that the body can serve as a model of intelligibility for understanding knowledge. Second, it compels us to consider this system of intelligibility as a consequence of a certain practice. Thus Glisson's originality is that he conceives knowledge as a medical practice. Consequently, the identification of a property of body, namely irritation, introduces an experimental

[28] "Nam potentiae sensitivae videntur esse participationes potentiae intellectivae. Et de his specialiter videtur loqui D Thomas, sed adhuc in his non est propia efficientia, sed est quasi quaedam causalitas exemplaris, nam cognitio sensitiva est quasi quaedam imitatio intellectivae; quomodo etiam opera partis vegetativae, inquantum fiunt quasi arte et intellectu, possunt dici imitationes intellectus." Francisco Suárez, *Commentaria una cum quaestionibus in libros aristotelis* de Anima (Sociedad de Estudios y Publicaciones, 1978), 2, §21.

[29] Oresme, *Questiones super Physicam*, 29ra–30ra.

model, and dissection is introduced into his reflection on knowledge. But if the experience of intrusion into the interior parts of bodies in dissection has consequences for the theory of knowledge, by the same token, it seems that another consequence for the understanding of the activity of the organs and the nerves is the possibility of understanding them as acting, and of doing so from an empiricist point of view. This is a stimulating suggestion for understanding the body in itself.

Acknowledgment

I would like to thank Heinrich von Staden for his careful reading and generous help.

6

The Organism-Mechanism Relationship

An Issue in the Leibniz-Stahl Controversy

FRANÇOIS DUCHESNEAU

1. Introduction

The organism-mechanism relationship formed one of the main issues in the Leibniz-Stahl controversy. After the publication of Georg Ernst Stahl's *Theoria medica vera* (1708),[1] Gottfried Wilhelm Leibniz sent a series of detailed remarks to the Halle physician who then replied. This first phase of the exchange was followed by a second series of objections on behalf of Leibniz, to which Stahl was to further reply by publishing the whole set of these writings in the *Negotium otiosum* (1720).[2] The general topics addressed in the exchange relate to contrasting views on the specific nature of living beings and on the relation of souls to organic bodies, but also to different methodological approaches to medicine as an art and a science. At the juncture of the metaphysical and epistemological theses that were debated, a clear difference emerges as to the nature of "organism" and the ways in which it should be accounted for. Stahl had been using this term since 1684. But the concept was brought to the fore of contemporary discussions around 1700. Raphaële Andrault has for instance shown that Nehemiah Grew made systematic use of it in his *Cosmologia sacra* (1701).[3] Leibniz himself introduced the term in a letter of his to Lady Masham in 1704. The first section of this chapter is devoted to tracing a significant trend that exerted direct influence

[1] Georg Ernst Stahl, *Theoria medica vera* (Halæ, DE: Literis Orphanotrophei, 1708).

[2] Stahl, *Negotium otiosum seu ΣΚΙΑΜΑΧΙΑ adversus positiones aliquas fundamentales Theoriæ veræ medicæ a Viro quodam celeberrimo* (Halæ, DE: Litteris et Impensis Orphanotrophei, 1720).

[3] See Raphaële Andrault, *La Vie et le vivant. Physiologie et métaphysique chez Spinoza et Leibniz*, Ph.D. thesis (Lyon, FR: École Normale Supérieure, 2010), 18–56.

on Leibniz and justified his appropriation of the concept of organism. In 1699, Hoffmann sent to Leibniz his recent dissertation *De natura morborum medicatrice mechanica*, in which he had developed original views concerning the organism. Leibniz replied approvingly and underlined significant similarities between the physiologist's arguments and the main concepts of his own theory of living beings. Hoffmann's arguments fed some of the theses Leibniz upheld and some of the criticisms he launched against Stahl concerning vital organization. In the following section, I address the confrontation between Leibniz and Stahl on accounting for the structures and functions of living beings through mechanism. The difference could prima facie be resolved into equivalence between mechanism and organism for Leibniz, and into an opposition between these for Stahl. This partial reading does not do justice to the basic theoretical differences in the theses that were at stake. I intend to show that these differences should be interpreted as connected to a common standpoint, namely the need to rely on "metaphysical presuppositions" to elucidate the issues relative to vital organization. On this account, the reference to Hoffmann's assessment of what would be methodologically required for a "rational medicine" provides collateral clarification. For Leibniz as for Stahl, the question that underlay their differences was that of the order of nature and the specific condition of living beings in nature. Within this metaphysical framework, the question was secondarily rephrased as a question about the nature of living beings as subjects of physiological properties and as agents of functional processes.

2. Hoffmann's Concept of Organism

Recent researches have shown that the concept of 'organism' was of use among physicians and natural philosophers from whom Leibniz borrowed the term while somehow shifting the meaning of it.[4] I personally was able to establish that the notion belongs to the doctrine of Friedrich Hoffmann, professor of medicine at Halle and himself a correspondent of Leibniz, witness the *Dissertatio inauguralis physico-medica de natura morborum medicatrice mechanica*, which Salomon Celarius defended on September 5, 1699.[5] The term had been previously used by Stahl in his *Dissertatio medica inauguralis de intestinis* (1684) to mean the property of an organ, at once formal and material, according to which

[4] We refer here, in spite of certain inaccuracies, to Tobias Cheung, "From the organism of a body to the body of an organism: occurrence and meaning of the word 'organism' from the seventeenth to the nineteenth century," *British Journal for the History of Science*, 39 (2006): 319–339.

[5] Friedrich Hoffmann, *Dissertatio inauguralis physico-medica de natura morborum medicatrice mechanica* (Halæ Magdeburgicæ, DE: Typis Christiani Henckelii, Acad. Typ. [1699]), without page numbers.

its composition and structure enable it to fulfill an instrumental end that can be identified with its function.[6] A body has this property when it is formed, administered, preserved, and repaired by a soul, which is a 'logos'-endowed agent. Both Hoffmann and Stahl were teaching medicine at the University of Halle. In this circumstance, we may probably infer that the term 'organism' was of use in the circle of influence of these two renowned physiologists, who were otherwise opponents in matters of method and theory.

Another important fact should be kept in mind. The controversy between Johann Christoph Sturm and Günther Christoph Schelhammer on the interpretation of the concept of nature had enticed Leibniz to take sides and write the *De ipsa natura*, which he had published in the *Acta eruditorum* in 1698.[7] Hoffmann refers to this polemics and seeks support from Leibniz's article for defending a conception of mechanism with which his concept of organism is directly linked. It is this circulation of arguments that provided the background for the exchange of letters between Leibniz and Hoffmann on the theme of 'rational medicine' (*medicina rationalis*) from 1699 on.

What does Hoffmann's *De natura morborum medicatrice mechanica* teach us? The object discussed is the concept of nature in its relation to the preservation of life and formation of organic bodies.[8] This Hippocratic theme needs be reconciled with a conception of phenomena that reduces them to the general physical order and submits them to a strict, rational approach.[9] Hoffmann argues that nature generally corresponds to an active principle in bodies and that, in humans and animals, this principle is the agent of vital actions that it constantly rules over and regulates.[10] Referring to the Sturm-Schelhammer controversy, he, like Leibniz, refuses Sturm's occasionalist hypothesis, which stressed the Cartesian distinction between *res extensa* and *mens* and thus deprived bodies of all inherent capacity of motion, of all autonomous dynamics. Hoffmann supports the view that bodies should be granted a real active force of their own; the existence of this force would justify that they be granted a corresponding passive force, which underpins their extension and capacity to receive motion.[11] Hoffmann goes to the point of setting an equation between substance and force. In this

[6] See Stahl, *Dissertatio medica inauguralis de intestinis, eorumque morbis ac symptomatibus cognoscendis et curandis* (Ienæ, DE: Typis Wertherii, 1684). R. Andrault (op. cit., 57–61) analyzes the four occurrences of the term *organismus* in this work, but stresses that the term seems to disappear later in Stahl, but to re-emerge with force in the *De mechanismi et organismi diversitate* (1706).

[7] See Gottfried Wilhelm Leibniz, *De ipsa natura sive de vi insita actionibusque Creaturarum, pro Dynamicis suis confirmandis illustrandisque*, in *Die philosophischen Schriften von Gottfried Wilhelm Leibniz* (Hildesheim, DE: Olms, 1965), GP IV, 504–516.

[8] Hoffmann, *De natura morborum*, §1.

[9] Ibid., introduction: "... *ad solidum rationis calculum omnia revocare....*"

[10] Ibid., §2.

[11] Ibid., §2.

case, it would be appropriate to conceive of bodies as 'extended forces' (*vires extensae*).

How do we get to organic bodies? These differ from inanimate bodies by a higher degree of structural composition. Hoffmann insists that the parts they are composed of are not similar and homogenous, but imply a building order that integrates the very diverse parts by giving them "figure, texture, and consistence."[12] This raises the question concerning the type of principle on which the operations of life depend. In critical fashion, Hoffmann here evokes the appeal to soul-like principles like plastick forces and archei, which he opposes by stressing that the nature of bodies, of whatever kind, consists only in particular and individual mechanisms.[13] Hoffmann specifies that mechanism, so understood, boils down to postulating that all effects of bodies, namely motions and operations, depend on physical principles acting in conformity with necessary laws and thus implying what he calls the "necessity of causes." Indeed, under the concept of natural mechanism, one may include physical and chemical processes that are not accomplished otherwise than through the action of necessary causes. But some mechanisms may more directly imply such a combination of parts that their organization would be a source of necessary motions under the notion of a coordination of effects fitting a determinate end: This end would be actualized in the very production of this mechanism and its ensuing operations. The example mentioned is that of the clock, which serves as an analogue of the more subtle and complex mechanisms that can yield vital motions. The definition Hoffmann offers for this particular modality of mechanism is indeed directly tied to a clause for its application to organized bodies:

> When several material causes are coordinated and disposed in such a way that effects arise from these that fit the idea of the craftsman who set to himself a determinate goal, it is still a mechanism, but more perfect, and, according to some, it is an organism, since this mechanism happens to exist in organic bodies.[14]

The reference to a more perfect mechanism resulting from divine design leads us undoubtedly close to the Leibnizian notion of "machines of nature" in contrast

[12] Ibid., §5.

[13] That was the argument developed by Robert Boyle in his *De ipsa natura sive libera in receptam naturæ notionem disquisitio* (Londini, UK: Typis H. Clark, Impensis Johannis Taylor, 1687), which had caused the controversy between Sturm and Schelhammer that Leibniz refers to in his own *De ipsa natura*.

[14] Hoffmann, *De natura morborum*, §5 : "...quando autem plures materiales causæ ita coordinatæ ac dispositæ sunt, ut effectus fluant, ideæ artificis, qui certum finem propositum habeat correspondentes, dicitur etiam mechanismus, sed perfectior, & a nonnullis organismus, quoniam in organicis corporibus existit."

to manmade machines.[15] Another passage shows in a remarkable way the correspondence between mechanism and organism:

> By this nature [of the Ancients] that preserves the body and heals it from illnesses we here understand not only an active principle that moves our machine, that is the motion that administers the parts and receives the ideas of external objects, but at once the whole mechanism or organism, that is the quantity and appropriate mixture of fluid parts, and the build-up and necessary configuration of solid parts, all this however without the rational soul.[16]

For Hoffmann, the mechanisms of the living body are presented as the effect of an organizing design issuing from the supreme intellect and operating on an extreme manifold of compounding parts, in terms of figure, size, and consistence: From this design and operation, there results an orderly combination of mixtures with a proportioned and harmonious organization of active and passive components. Consequently, such an arrangement generates motions qualified as vital that enable the mixed components integrated into the organic body to be preserved and to last for a while. To these vital motions the nutritive functions are linked, as well as the sensitive and motile ones; the motions implied in the latter "tend and conspire" to execute actions conformable to commands of the mind.[17] The difference between these divine machines and artificial ones essentially consists in the infinitely diversified modes of action of an infinite power compared to the limited architectonic power of humans, a power that acts by simple means at the surface of bodies without being able to structure their inner composition to infinity. Hence, a feature of the divinely achieved products is that living bodies are animated by an internal and autonomous principle of motion whose action extends equally to all constitutive parts; by contrast, in the pneumatic as well as hydraulic automata, motion arises from essentially external sources.

[15] On Leibniz's notion of machine of nature, see Michel Fichant, "Les Machines de la nature," *Studia Leibnitiana* 35 (2003): 1–28; Justin Smith, *Divine Machines: Leibniz and the Sciences of Life* (Princeton, NJ: Princeton University Press, 2011).

[16] Hoffmann, *De natura morborum*, §8 : "...subintelligimus per naturam, quando dicitur ab antiquissimis medicis, corporis conservatrix & morborum medicatrix, hic non tantum principium machinam nostram movens activum, sive motus in partibus celebrans & externorum obiectorum ideas recipiens, sed simul comprehendimus totum mechanismum seu organismum, id est, partium fluidarum legitimam quantitatem, temperiem, solidarum debitam constructionem ac configurationem, exclusa tamen anima rationali."

[17] Ibid., §6: "...qui motus in animantium corporibus appellatur vitalis sive ipsa vita, quæ ad conservationem & durationem principaliter corporis mixti & structi, postea etiam ad nutritionem, item sensationem & motus, mentis imperio obsecundantes, tendit atque conspirat."

But the difference par excellence relates to the power of reproduction of natural machines by seminal power, a power entirely lacking in artificial machines.[18] Presuming that there has been a primeval creation of organic dispositions, Hoffmann raises the question of how the reproductive power living beings are endowed with is actualized. He excludes the hypothesis that God would form each animal or plant in particular in the sequential order of natural generations. He rather opts to attribute to each organic body a sui generis power for reproducing itself and perpetuating its species, provided God generally concurs to the preservation of the created world.[19] This power resides in the 'seminal principle' (*principium seminale*) that forms "the prime elements and rudiments of the body" (*prima lineamenta & rudimenta corporis*). Does this seminal power act with intelligence—as suggested by Stahl's attributing a "logos" to the living being—or without any cognitive capacity? This metaphysical question is set aside for the benefit of a hypothesis that fares as an extension of the mechanist model. It suffices to postulate a most subtle and active principle that is apt to receive impressions and dynamic determinations and to produce correlated effects independently of all volition. In an analogical way, the seed generates such or such an individual without the direct intervention of any soul-like agents. Hoffmann seems to suggest that specific motile dispositions embodied in extremely tiny organic arrangements can generate morphogenetic, as well as sensitive-motile processes in accordance with some psychic determinations; these are equally characteristic of the compound substance that forms the living being, at once soul and organic body, and grounds the correspondence between perceptions and functional sequences of physicochemical operations. Thus Hoffmann argues that everything happens as if the soul had the power to act on the body, as shown by the passions, and reciprocally, as if the states of the body and notably the dispositions of its humors might affect the mind and modify it.[20] Hence one can say that the living being results from the substantial integration of a soul and organic body—an integration whose nexus is presupposed although it escapes our intellective capacities.[21]

At the same time, however, the structural and functional autonomy of the organic machine is set as a condition for the analysis of all vital operations, including those that relate to sensibility-motility:

> Life, nutrition, the processes of secretion and excretion, the circulation of the blood and humors, the impressions from external things on the sense

[18] Ibid., §7.
[19] Ibid., §6.
[20] Ibid., §8.
[21] Ibid., §8.

> organs themselves, the involuntary motions of parts and muscles, are merely mechanically achieved and accomplished by physical corporeal causes, without necessary assistance, concourse or operation of any principle of free action. This is why, concerning this mechanism, we declare that it is at the highest degree a work of wisdom, order and intelligence, arranged by God with supreme intelligence and endowed with instruments and means such that it possesses sufficient forces for preserving the machine and defending it against putrefaction and the destruction of its combination of mixtures, but also for repelling and avoiding all internal lesion that would tend to ruin the machine: and however all this is done without any intelligence, by reasons internal and proper to the body.[22]

The remainder of the dissertation concerns the implications of this conception of vital organization for pathology. Generally, Hoffmann seems to adopt a Stahlian vision on the latent, imminent corruptibility of the combination of mixtures that the organic body represents. But, contrary to Stahl, he systematically relates preservation against this corruption not to the intellective and volitional infra-conscious capacities of the soul, as the immaterial agent of salutary motile dispositions, but to the very organization of the organic machine and the motile dispositions arising from it. The organic body, because of its original formation, would be endowed with means "sufficient for its preserving and defending itself mechanically against corruption and death, since what is noxious and adverse to the combination of mixture and vital motion provokes violent and extraordinary flows of spirits in the muscular parts for correcting or eliminating the pathogenic matters."[23] Organic dispositions of a mechanical, hydraulic, or static type bring about the circulatory movement of organic fluids and prevent by the same means their prompt decomposition. This movement of the blood and/or spirits appears as a natural instrumental means to respond to the deleterious

[22] Ibid., §8: "Attamen vita, nutritio, actus secretionum, excretionum, sanguinis & humorum circulatio, impressiones rerum externarum in ipsa sensoria, motus partium & musculorum involuntarii, mere mechanice a causis corporeis physicis conficiuntur & absolvuntur citra ullum necessarium auxilium, concursum, vel operam, principii cuiusdam libere agentis. De hoc itaque mechanismo pronunciamus, esse ipsum opus summe sapientiæ, ordinis, intelligentiæ, & ita a Deo summo intelligente dispositum ac instrumentis & mediis munitum, ut non tantum machinam conservare, a putredine & destructione mixtionis suæ defendere, sed & læsiones quasvis internas, perniciem machinæ intententes, propulsare atque propellere, sufficientibus viribus instructum sit, quamvis hæc omnia fiant sine quadam intelligentia, ratione ipsius corporis interna & propria."

[23] Ibid., §14: "Tantis nempe præsidiis instruxit divinum Numen corpus nostrum, ut sufficiat sæpe numero sibi ipsi se tueri & defendere mechanicis rationibus a corruptione & morte, dum, uti dictum, illud ipsum quod perniciale & infensum mixturæ & motui vitali est, proritat violentum & extraordinarium spirituum in partes musculosas influxum, qui deinde serviat causæ morbificæ lædenti vel corrigendæ vel exterminandæ."

effects generated by pathological factors, themselves interpreted according to the categories of physicochemical mechanism.

On this perspective, it is not surprising that in his letter to Hoffmann of September 27, 1699, Leibniz acknowledges receipt of the dissertations the Halle physician had sent him and comments with approval on the main theses to be found in them and especially in the *De natura morborum medicatrice mechanica*, but with adjustments so that they fit his own conceptions of living beings and organic bodies. As for the pattern of 'rational medicine' that Hoffmann brings forward in conclusion of his dissertation, Leibniz feels that it converges with the methodological theses he favors and whose relevance he defends against Stahl.[24]

3. The Opposition between Leibniz and Stahl on the Theory of Organism

Stahl published his *Disquisitio de mechanismi et organismi diversitate* in 1706 and republished it as the first introductory essay of his *Theoria medica vera* in 1708. This is a clear indication that he stresses the importance of the new concept of 'organism.' On the other hand, he seems to set it in radical opposition to 'mechanism,' which is in a way an unprecedented move. We can probably summarize what Stahl means by organism by pointing out that the term qualifies individuals that belong to natural species. These entities are destined to produce an end for which all their structures and actions, however they may be decomposed, appear as necessary means and thus as subordinate ends.[25] Moreover, the beings forming those species only exist and subsist for achieving this essential end. And such an end can only emanate from an agent possessing intentional aims. In the natural order, the agents here considered reside within the bodies they form, organize, and animate, whose integrity and formal persistence they maintain through the almost infinite sequences of modifications those bodies undergo because of their materiality. This is indeed where 'mechanism' comes into play. This term means all the processes of modification affecting bodies in their materiality, and thus lacking any final determination.[26] An important feature of mechanism thus characterized consists in its separation from any efficient relationship that would suppose that it be determined by the imposition of a form or end. The world of phenomena reduced to mechanism would consist in mere physical and chemical facts arising from blind necessity. If a certain order can be thus generated, it must

[24] See François Duchesneau, *Leibniz. Le vivant et l'organisme* (Paris: Vrin, 2010), 155–158.

[25] See Stahl, *Disquisitio*, §§13–14, in *Theoria medica vera* (Halæ, DE: Impensis Orphanotrophei, 1737), 7.

[26] See ibid., §31, 13.

be fortuitous and characterized as an effect of chance (*casus*), like the hills that wind forms by moving about multitudes of grains of sand, without any evident final destination, without any goal set beforehand for ordering the process.[27] Following Democritus and Epicurus, the moderns have neglected organism by lack of understanding it; they have focused on mechanism alone and attempted to draw hypothetical laws ruling over it and grant them universal application.

But some natural productions cannot be reduced to those laws since their inner order can only result from an organic, that is, an instrumental, determination. This determination corresponds to a reciprocal relation between ends and means, and such a relation would be achieved at every integrative level in the formation and operations of the living organic being.[28] Therefore the instrumental relationship that ties together the body's structural and functional components provides the main defining feature of what is to be understood by 'organism,' and accordingly determines the subordination of mechanism to organism.[29] One should here understand that the living being depends on a functional integration, and not on a mere coordination, of mechanisms: Such coordination could achieve a seemingly orderly result, but this result would not have formed an intrinsic goal for the processes involved. Stahl gives as an example the mechanical framing of river banks and the production of salts from reactions between acids and alkalis. In contrast, the organism that characterizes substantial entities, as well as physical operations, necessarily implies mechanical devices, but the constitutive raison d'être that causes the formation, integration, and efficiency of these surpasses mechanism.[30] Stahl refers to the example of artificial machines, such as the clock, to underline that we can get analogous combinations of mechanisms in nature, but that these can only be considered as organisms when they achieve ends for which the agent who is responsible for their formation and operations has conceived and organized them. Such an agent is external to the artificial machine while it is an intrinsic component of the natural machine.

These are the conditions and roles of the soul in Stahl's physiology. This soul only exists and acts to provide instrumental finality to the living body by generating and regulating the motions affecting corporeal structures.[31] Reciprocally,

[27] See ibid., §18, 8–9.

[28] Ibid., §34, 14: "Tanto maior autem est evidentia huiusmodi etiam reciprocæ indolis *efficientis & finis*, quando efficiens illud non solum ita directe ad illum tendit, sed etiam de reliquo, neque per se exsistere aut subsistere apparet, nisi quatenus ad illum finem agit atque tendit, neque effectus directus & primarius ulli alteri rei æque quadrare, imo vel solum directe subservire apparet, quam solum ut illud efficiens energiam suam non magis exerceat quam exhibeat, adeoque vere & simpliciter effectus de sua causa testetur."

[29] See ibid., §36, 14–15.

[30] See ibid., §39, 16.

[31] See in particular ibid., §81, 33–34.

the living body essentially appears as the soul's instrument or organ, and this determines the modes of formation and functional processes that express the organism.[32] There are two fundamental features of vital motions: (1) Because of the corruptibility of the aggregate of mixtures forming the living body, Stahl gives the soul a causal role in the preservation and restoration of the living being's integrity. Everything occurs as if the soul formed a force of opposition to the disintegration and death of the living being, as if the mechanisms left alone without this active direction by a foreign principle were doomed to decomposition. However, Stahl's position in this instance does not involve a strict anti-mechanism, contrary to some standard interpretations. For the conservation, as well as restoration, of living beings is rather presented as the result of the integrated mechanisms themselves, but only as long as the soul has provided for their formation and maintenance.[33] (2) To explain physiological, as well as pathological, phenomena, Stahl constantly relies on the 'organic nexus' (*organicus nexus*) between the soul and the body as an instrument.[34] While stating that this 'nexus' is not directly accessible to the intellect, he reorients the effort to understand it toward the 'synergies' that express the "mechanical-organic nexus for the preservation, defense and restoration of the body."[35]

Faced with these neatly expressed hypotheses on the opposition and complementarity between organism and mechanism, Leibniz begins his critique by recalling some elements of his natural philosophy. In this way, he develops a partly different and contrasted notion of organism, grounded in his concept of machine of nature. But, as usual with him, he undertakes to integrate with his theory arguments that are more directly inspired by Stahl's assertions on the structural and functional organization of the living.

The preamble to the *Animadversiones* precisely sets up the framework of natural philosophy that Leibniz has inferred from previously acquired arguments. Concerning the organism issue, certain points are brought to the fore, and first of all, the autonomy of order of physical phenomena. For the application of the principle of sufficient reason implies that "everything that occurs in matter flows from the previous state of it according to the laws of change."[36] As for these laws, they form the kingpin of all explanation by way of mechanism. In the order of

[32] See ibid., §84, 35.

[33] See ibid., §68, 28.

[34] Ibid., §91, 38.

[35] Ibid., §97, 42: "Hac demum solida considerandi methodo vere dilucescet, quid activum in corpore, utile imo necessarium, pulchro ordine atque nexu vere mechanico-organico ad conservationem, vindicationem, & restaurationem corporis fiat atque faciat...."

[36] Leibniz, *Animadversiones circa assertiones aliquas Theoriæ Medicæ veræ Clar. Stahlii, in Opera Omnia* (Genevæ, CH: Apud Fratres de Tournes, 1768), Dutens II–2, 131: "...quicquid in materia evenit, ex præcendenti materiæ statu, per leges mutationum oriri."

phenomena, nothing outrides mechanism. Indeed, the order and laws of phenomena require to be causally grounded in the theory of monads and the laws ruling over their sequences of inner changes; and we know of the principle of harmonic correspondence that warrants the correlation between the sequences of monadic perceptions-appetitions on the one hand and the sequences of phenomena in the organic body on the other hand, because of original divine preformation. The key argument in Leibniz's interpretation is summarized in a significant statement of the preamble:

> ... in the organic body of the living being which the soul presides over as a particular ruling agent, even if the source of all action is in the soul, nothing occurs however contrary to the laws of bodies, as nothing happens in the soul if not according to its proper laws, although the source of its passion comes from matter.[37]

In the exchange that follows, Leibniz often stresses, in support of this thesis, that it is impossible to conceive how the geometrical and mechanical properties of bodies may arise from the cognitive and conative properties of monads and vice versa. To borrow Stahl's terminology, it is the 'organic nexus' that is properly unintelligible in its own reality and requires that it be substituted with the expressive correspondence between the corporeal and monadic instances. It is because of original preformation alone that the correlation is such that "the soul is the essential representation of the body and the body the essential instrument of the soul."[38] And the principle of expressive correspondence applies to the entire set of phenomena, as well as to the entire set of monadic substrates. Hence the impossibility of restricting the final or instrumental determination to a single type of natural entities: The finality and causal efficiency are universally distributed in both spheres of reality. Thus one can use teleology to explicate the order relations underpinning those phenomena that are specific to inorganic bodies, optical or mechanical processes, for instance. Consequently, instrumental finality does not appear anymore as the exclusive mode of accomplishment of organic phenomena, but rather as one of the two possible ways for explaining them, the other consisting in displaying the sequences of mechanical effects by which physiological operations are achieved; this way is indeed deeper and more difficult than resorting to teleology, as already mentioned

[37] Ibid., Dutens II–2, 133: "... ita in corpore organico viventis, cui anima tanquam rector peculiaris præest, etsi omnis actionum fons sit in anima, nihil tamen fit præter corporum leges ; uti vicissim nihil in anima, nisi per proprias leges, oritur, etsi fons passionum ejus ex materia oritur."

[38] Ibid., Dutens II–2, 133: "... ut anima sit essentiale corporis repræsentativum, & ut corpus sit essentiale animæ instrumentum."

in section 22 of the *Discours de métaphysique*.[39] The same is expressed in the preamble to the *Animadversiones*: "Even if the effects arise from the inner motions and structure of the machine, as these internal features are less known to us, they can be guessed more easily by the ends than by the mechanisms."[40]

In these conditions, it is easy to understand that Leibniz opposes the presumed real difference between organism and mechanism. He evokes two main objections: The first consists in postulating that organism is only a higher degree of mechanism since it characterizes the inner order of organic bodies, that is, machines of nature that are divine productions in contrast to manmade machines.[41] The important point is to acknowledge a principle of composition of structures and processes that unfolds infinitely within the machine and whose formal unity, a source for integrating its inner variety, resides in an entelechy alone. But analyzing these structures and processes in their own right would not require resorting to other modes of explanation than those of mechanism. This would justify the arguments brought forward in the following article of the *Anidmadversiones*: the rejection of any element foreign to mechanism in the organism on the one hand; the assertion of an absolute domination of mechanical reasons in the field of organic phenomena on the other. There is then an allusion to plastick natures and other principles of the type of Stahl's physiological soul that would own the capacity for directly intervening in the phenomenal order to modify it: Such organizing agents heterogeneous to mechanisms could not bypass the principle of sufficient reason. Indeed, Stahl had argued that the agent in the organism grounds its action in the mechanical devices composing the living body in order to fulfill the instrumental end forming its destination, but Leibniz does not believe that this concession suffices for harmonizing organism and mechanism in the unity of a concrete living being. The mechanisms involved in structures and processes do not only constitute this being materially, but also formally: "It should be added that the organism is nothing else formally than a mechanism, but more subtle and divine, since everything in nature must occur mechanically."[42]

[39] Leibniz, *Sämtliche Schriften und Briefe* (Berlin: Akademie Verlag, 1999), A VI 4, 1565: "Cependant je trouve que la voye des causes efficientes, qui est plus profonde en effect, et en quelque sorte plus immediate et à priori, est en recompense assez difficile, quand on vient au détail, et je croy, que nos philosophes le plus souvent en sont encor bien eloignés. Mais la voye des finales est plus aisée, et ne laisse pas de servir souvent à deviner des verités importantes et utiles qu'on seroit bien long temps à chercher par cette autre route plus physique."

[40] Leibniz, *Animadversiones*, Dutens II–2, 135: "Etsi enim effecta oriantur ex intestinis motibus structuraque machinæ : tamen quia horum interiora ignota sunt, facilius ex finibus, quam ex mechanismo, divinari possunt."

[41] See ibid., §2, Dutens II–2, 136.

[42] *Exceptiones*, §2, Dutens II–2, 144: "... addi enim debet, organismum nihil aliud esse formaliter, quam mechanismum, etsi exquisitiorem & diviniorem, quia omnia in natura fieri debent mechanice."

If we refer to articles 8 and 9 of the *Animadversiones,* we get a further paradigmatic differentiation about organism according to each author. The issue here is about what justifies acknowledging the organic body as living. For Leibniz, the determining condition of life consists in the capacity for perception and appetition, that is, in the formal feature of the leading monad associated with the organic body. For Stahl, life resides in the very order of phenomena and is revealed by the antagonistic action of the soul, an agent of final determination, in relation to the material dispositions that the whole set of organs of the body represents: These material dispositions, which belong to a relatively unstable chemical aggregate, are to dissolve when left to themselves, that is, when undergoing the processes of a chemistry pertaining to inorganic bodies. In his *Enodationes,* Stahl, besides protesting against Leibniz's interpretation that the function of life would then be reduced to the mere preservation of the body by itself, links this function to the soul, but insofar as this agent's behavior is determined by the state of the body and by the finality to be implemented in the determination, degree, proportion, and duration of the specific vital motions. However, because the soul's relation to its organic effects is ruled over by intentionality and implies a relative contingence of the nexus between psychological causes and organic effects, a perverted intention may occur whose result, unfolding in functional but insufficiently adapted motions, would be pathological, and even lethal.[43] Stahl points to the ambiguity affecting the notion of force of vegetation: Among many interpreters, he says, this term is used at once to designate life and the physiological processes of nutrition and body repair identified as effects of the organic arrangement. It should be appropriate to distinguish between, on the one hand, the mechanisms involved in those processes, which can last for a while, notably in adult beings before decomposition starts, and on the other hand, "the directly conservative act" (*actus directe conservatorius*)[44] of the principle of animation and its subordinate organic operations: circulation of humors, secretions, excretions, and other regulative processes.

Leibniz transposes Stahl's thesis with a twofold reduction: (1) He attributes the functions of preservation, restoration, and reproduction to a vegetative force that arises from the very structures of the organic body (at least this would be the object for a possible reduction to the material conditions of the organic device understood in terms of motions and micro-motions); and in this context, Leibniz draws an analogy between this organic device and a flame. (2) Leibniz radically reduces the role of the soul in these processes by only focusing on effects of regulated concordance without direct causation by the animating principle on the system of corporeal processes by which life manifests itself in express and specific fashion. But, at the same time, he formally attributes life to the soul as the

[43] Stahl, *Enodationes,* §8, in *Negotium otiosum,* 45.

[44] Ibid., §9, 46.

subject of perception and appetition. Life thus becomes a psychological equivalent of a corporeal mechanism that suffices for maintaining, accomplishing, and reproducing the functional motions by which vitality is organically expressed. Article 9 of the *Exceptiones* underlines the epistemic priority of the mechanist explanation of phenomena occurring in the organic body. It is the inadequacy of this mode of explanation that should have been proven, which Stahl did not succeed in doing. On the other hand, Leibniz rejects the presumed absence of analogy between the maintenance of the flame and the functional autonomy of the machine of nature. He establishes the strict determination of both processes by external physical causes; the inner order of the physiological operations is thus integrated into the vaster field of external corporeal conditions warranting the constant flow of micro-processes. Whereas Stahl separates the sphere of the living individual from the more encompassing one of inorganic corporeal phenomena and specifies the causal autonomy of the agent of life, Leibniz extends the sphere of organism to all corporeal phenomena; throughout the order of nature, organic phenomena can be analyzed according to the same patterns of mechanical and dynamic efficiency as inorganic phenomena; in parallel, however, the formal raisons d'être of this general organized system are referred back to the intrinsic order of monads, including animal souls.

In the *Paraenesis ad aliena a medica doctrina arcendum*, another prefatory essay to the *Theoria medica vera*, Stahl had stated in aporetic fashion that vital motions in their integrative determination would not depend "on a just proportion of the matter and corporeal organs."[45] He had added the supposition that if this proportion were held to self-preserve, vital motions would continue by themselves without interruption and, if they happened by chance to have been altered, they could indeed reconstitute themselves. But a posteriori reasons to the contrary could be evoked: In particular, disorders in vital motions tend to occur without regular physical causes; and those motions may be accelerated or slowed down short of any proportion with the specific organic dispositions wherein they are accomplished. Hence Stahl's recourse to the affections and habits of the soul as sources for these altered processes. Leibniz counters this conception, which he considers destructive for all forms of mechanist interpretation of the vital economy. He corrects it by adjoining to it the hypothesis that subtle motions of the imperceptible parts enter the composition of the organic body, without which it would be impossible to account for the observable physiological, as well as pathological, processes. Appetites and other affections of the soul that seem to directly generate organic alterations would thus only appear as the expressions of concurring phenomena. The requirement of looking for intelligible sufficient reasons

[45] Stahl, *Parænesis*, §33, in *Theoria medica vera*, 59: ". . . a materiæ & corporeorum organorum justa proportione."

for corporeal processes entails appealing to the specific organization, or 'organism,' of the machines of nature. What should one conclude from this, but that "the more distinct the machine is, the more manifest it renders the craftiness of a divine structure"?[46] In the *Exceptiones,* Leibniz argues that if in a certain measure, the principle of motion can be distinguished from the vital motion it generates, it remains that this principle belongs to the material order and organic dispositions, since moving a body of whatever size and figure implies an impulse from another body itself in motion. The pure genesis of motion in the order of phenomena by some non-corporeal cause would break with the intelligibility of causal sequences. Indeed, the organic body can undergo transformations by addition or subtraction of force, but this occurs through an exchange of power effects between the organic body and external bodies, as well as within the integrative system that the organic body forms.[47] The analogy that reappears here is that of a machine that would receive, inwardly distribute, and spend energy following self-regulatory processes.

Another innovative aspect of the theory of organism stems from the role and importance of chemistry for providing mechanist models. For Leibniz, these models would foster the explanation of organic phenomena; for Stahl, they must be outridden in physiological analysis.[48]

At the center of his model for physiological activity, Leibniz reasserts against Stahl that the agent of impetus is related to the power of the micro-components of the vital fluids interacting with the micro-structures of the organic solids. Hence certain considerations concerning what may be termed a vital chemistry. So is it with the remarks about the action of volatile salts of urine[49] that serve as anticoagulants of blood in vitro; this reaction taking place outside of the organic body does not appear to Stahl to help account for any physiological process or foster any therapeutic use. In article 24 of *Animadversiones,*[50] Leibniz praises this type of experience and suggests that at a different degree of intensity and with much lesser quantities, it could serve as an analytic tool for understanding functions and affording treatments, provided that the other chemical agents jointly intervening with these salts have been identified. On the same line, Stahl would question the importance of altering drugs, and even that there could be anything of the sort. In cases of organic dysfunction, he would rather limit the possibility of intervening to the evacuation of peccant matters, either through the body's normal operations or by external interventions (bloodletting, purgation, etc.).

[46] Leibniz, *Animadversiones,* §15, Dutens II–2, 139: "...quanto hæc animalis machina præstat exquisitius, tanto magis divinæ structuræ conspicuum artificium esse."

[47] See *Exceptiones,* §15, Dutens II–2, 149–150.

[48] See Duchesneau, *Leibniz. Le vivant et l'organisme,* 144–153, for an analysis of this aspect of the controversy.

[49] Ammonium carbonate.

[50] Leibniz, *Animadversiones,* §24, Dutens II–2, 141.

In article 25,[51] Leibniz objects to this statement because of the febrifuge action of the powder of cinchona, as well as the effects of drugs with opium or arsenic components. More generally, he underlines that physicochemical alteration is everywhere present in the normal functioning of organs and hence in means of intervention to restore health. Even evacuant agents produce effects only by altering vital processes, and all these forms of alteration can be analyzed according to a pattern of interactions between organic fluids and solids on the one hand, and substances introduced into the body for the sake of treatment or experimentation on the other. The crucial notion of alteration points to a program fostering multiple analyses of the dynamic interaction between the various solid and fluid masses constituting organs and chemical compounds of mineral, vegetal, or animal origin. In a general way, it is through chemistry that the inner mechanisms of organic bodies that determine and regulate their functional operations can be unveiled. These mechanisms, by their integrative and regulative action, represent at the phenomenal level the infinite integration of the living as compound substances. Leibniz conceived the chemistry of his time as essentially empirical, focusing on the properties and reactions of similar or quasi-similar bodies, that is, bodies whose parts are quite homogenous—meaning inorganic bodies that in turn will form organic ones. But this science may undergo two significant shifts that would enhance its capacity for explaining vital phenomena. It can correlate its more general statements, which apply at once to inorganic and organic processes, with the results of microscopic anatomy concerning combinations of parts; this first elaboration entails a broadening of the sphere of empirical knowledge. But on the other hand, the data provided by these micro-anatomical and chemical analyses could be subordinated to a theoretical framework that would conform to the *mathesis*, by resorting to one form or other of mechanist model; in this way, the sphere of theoretical explanations could and should be expanded.

Hence the expected coming of age of physics in its application to machines of nature.[52] Thus Leibniz envisions the setting up of a 'special physics' (*physica specialis*) for unveiling the operations of imperceptible parts of organic bodies in their relationship to vital functions, notably concerning the fibrils of nerves and membranes. But these reasons must be borrowed in large part from a mechanist chemistry bearing on organic micro-processes. Thus it is appropriate to interpret the definition of the animal body as a *machina hydraulico-pneumatico-pyria* as meaning in the later period a combination of dynamic micro-processes whose agents of *impetus* should be identified through chemical reactions matching specific functional properties.[53] Thus Leibniz earmarks the action of the circulatory

[51] Ibid., §25, Dutens II–2, 141–142.

[52] See *Exceptioness*, §11, Dutens II–2, 148.

[53] See *Animadversiones*, §13, Dutens II–2, 149.

network, which at that time combined blood, nerve, and lymphatic vessels, in the intense, rapid, and efficient transformation of particles of nutritive mixtures for the various specialized organic functions; these transformed particles are subject to dissipation, involution, and secretion—a technical terminology that symbolizes the various modalities of organic assimilation and disaggregation.[54] This would form a convincing illustration of the Leibnizian equation between organism and mechanism resulting from the inherent teleological ordering of the implied physicochemical structures and processes.

4. Conclusion

Leibniz, as well as Stahl, resorts to the new notion of 'organism,' which at the turn of the 18th century serves to characterize the structural and functional arrangement of plants and animals. Undoubtedly, the practice of subtle anatomy (*anatomia subtilis*) has contributed to reveal the extreme structural complexity of the integrative parts of organic individualities. How does this combination of structures, this skillful inner geometry, this architectonics of living bodies condition the processes by which vital functions are accomplished? What connection is there between the organic dispositions of the living beings thus described and the principles of animation that the ancient physiology supposed? Faced with this twofold issue, Friedrich Hoffmann, one of the major theoreticians of the new physiology, had developed a conception of 'organism' as a special mechanism proper to organic bodies. The dissertation *De natura morborum medicatrice mechanica* (1699) was approved by Leibniz who found in it a physiological model of the living being that appeared sufficiently compatible with the mainframe of his own theory of machines of nature. And he relied on this component of the program for a 'rational medicine' when he wrote his remarks on Stahl's *Theoria medica vera* (1708) and criticized the Stahlian concept of organism. For Stahl, as for Leibniz, the field of application of the notion of organism is conditioned by a certain relation to mechanism as a general principle for the analysis of organic, as well as inorganic, phenomena. Indeed, Leibniz favors the equation between organism and mechanism, while Stahl tends to oppose these two modes of determination to each other. Beyond this main difference, there are several other consequences to be drawn from the divergent views exposed in the Leibniz-Stahl controversy. Among these, divergent conceptions about the object of physiology afford significant alternative frameworks for building theoretical accounts of the structural and functional organization of organic bodies.

[54] See *Animadversiones*, §14, Dutens II–2, 149.

PART THREE

THE GENERATION OF LIVING BEINGS

7

Material Causes and Incomplete Entities in Gallego de la Serna's Theory of Animal Generation

ANDREAS BLANK

1. Introduction

The contribution of late Scholastic thought to the development of the early modern life sciences has not received much attention in the recent growth of interest in the history of the early modern life sciences and its relation to philosophy. One exception is Gómez Pereira (1500–c. 1558), whose theory of animals as inanimate machines strikingly anticipates some traits of Descartes's natural philosophy.[1] But Pereira was by no means the only innovative late Scholastic contributor to issues in the life sciences. Some quite different but no less interesting material concerning animal generation is to be found in the work of Juan Gallego de la Serna, royal physician to the Spanish kings Philip III and Philip IV.[2] Although Gallego accepted the traditional view that animals possess an animal soul, he nevertheless attacked two other widely accepted views: (1) the view that animal seeds are animated,[3] and

[1] See Gómez Pereira, *Antoniana Margarita* (Medina del Campo, IT: n.p., 1554). On Descartes's relation to Pereira, see Gabriel Sanhueza, *La pensée biologique de Descartes dans ses rapports avec la philosophie scholastique: Le cas Gómez Pereira* (Paris: L'Harmattan, 1997).

[2] Gallego's biographical data are not recorded by any of the specialized libraries holding his writings. So, one has to go by the dates of his two major publications: Juan Gallego de la Serna, *Opera physica, medica, ethica, quinque tractatibus comprehensa* (Lyon, FR: Iacobus & Petrus Frost, 1634); Juan Gallego de la Serna, *De naturali animarum origine* (Brussels, BE: Franciscus Vivienus, 1640). The latter work was edited posthumously; Gallego's preface is dated May 1638. Both works are available online at googlebooks.com.

[3] This view was accepted, among others, by Girolamo Cardano, Julius Caesar Scaliger, Jacopo Zabarella, and Daniel Sennert; for detailed expositions, see Thomas Feyens, *De formatrice foetus* (Antwerp, BE: Gulielmus a Tongris, 1620) (available online at googlebooks.com), 44–54; Gallego, *Opera*, 64–68.

(2) the alternative view that animal seeds, even if not animated, possess active potencies sufficient for the development of animal souls.[4] Thus, even if Gallego's view of the nature of animals is much less radical than those of Gomez and Descartes, his view of the nature of animal seeds diverges radically from what almost everyone else believed in his time.

For someone who accepted a Galenic two-seed theory of animal generation, as did Gallego, both the view that animal seeds are animated and the view that they possess active potencies sufficient for the development of animal souls pose a deep problem. Suppose that seed souls are immaterial entities (an assumption made, e.g., by Gallego's prominent contemporary Daniel Sennert). Such a supposition seems to present no problem concerning the persistence of these souls across the process of conception. However, there seems to be one soul too many in the fetus, and some (probably implausible) argument has to be provided to explain why only one of them functions as the soul of the nascent animal. Alternatively, suppose that animal souls are features of the material organization of seeds, such as the temperament of primary qualities. On such a supposition, it seems conceivable that once the matter of the two seeds is all mixed together and then mixed with menstrual blood, the temperament of this mixture would function as the soul of the nascent animal. Thus, there would not be too many souls in the fetus. However, there seems to be a problem regarding how the temperaments of the seeds could be understood as persisting in the temperament of the mixture of seminal matter with menstrual blood. It would rather seem that the temperaments of the seeds have ceased to exist, and an entirely new temperament has come into being.

Indeed, the view that animal seeds are nothing but material causes of animal generation is not entirely unprecedented. Gallego points to two predecessors whose ideas he partly accepts. One of them is Duns Scotus (1265/66–1308); the other is Gallego's contemporary Thomas Feyens (1567–1631). Scotus's influence on other areas of early modern thought such as philosophical theology is well known. Still, his influence on early modern theories of biological reproduction has not garnered much attention from commentators. This is regrettable because Scotus presents some intriguing arguments to show why it does not make sense to suppose that substantial forms of seeds can persist across the process of conception. Accordingly, he suggests that animal seeds are not animated by vegetative or sensitive souls but rather possess substantial forms that function as transitional, "intermediary" forms that are neither souls nor persist in the fetus. Feyens agrees that the active powers of seeds cannot be understood

[4] This view was accepted, among others, by Aquinas, the Coimbra Commentators, Jean Fernel, Laurent Joubert, and Francesco Piccolomini; for detailed expositions, see Feyens, *De formatrice foetus*, 88–91; Gallego, *Opera*, 22, 33.

as powers identical with the active powers found in the fetus and adds to Scotus's metaphysical arguments some arguments drawing on the medical tradition. Gallego finds this line of thought attractive; in section 2, I outline the arguments that Gallego shares with Scotus and Feyens.

Yet, neither Scotus nor Feyens presents an explanation of the origin of vegetative and sensitive souls in terms of natural processes. Scotus explicitly invokes divine agency to fill this explanatory gap, while Feyens does not offer any explanation. It is this explanatory gap that Gallego tries to fill without having recourse to God as a primary cause. To be sure, Gallego accepts the view that God should be understood as a concurrent cause of all natural events that preserves the existence of the primary causes.[5] But he suggests that two ideas can explicate the nature of primary causes relevant for sexual reproduction: (1) the theory of matter and form as incomplete entities developed by the Spanish Jesuit Francisco Suárez (1548–1617); and (2) the Galenic view that the uterus possesses a formative faculty of its own. In section 3, I explore how Gallego modifies the theory of incomplete entities to explain how animal souls can be "educed" from the potencies of matter. In section 4, I examine how Gallego uses the idea of a formative power of the uterus to fill out the abstract framework of a theory of incomplete entities with some concrete physiological mechanisms.

2. Seeds as Material Causes

Gallego is a highly eclectic thinker whose own, quite innovative proposals are made against the background of the acceptance of numerous elements in the thought of his predecessors. This is why this section and parts of the two following sections trace not only Gallego's own views but also the views of others that he builds upon. Gallego makes it clear that what is interesting about Scotus's position is the thesis that an animal seed is neither a "primary active cause of substantial formation" nor an "instrumental active cause of substantial formation," but rather only a "material cause of substantial formation" and an "efficient principle of preliminary dispositions."[6] This characterization is somewhat confusing because it both ascribes and denies some active potency to the animal seed. The appearance of confusion, however, quickly dissolves once one realizes that the active potencies at stake are distinguished by the differing natures of their effects. Some active potencies are those that lead to the generation of substantial forms. It is this kind of active potencies—be it primary (i.e., active due to its own nature) or instrumental (i.e., active by virtue of the agency of

[5] Gallego, *Opera*, 133.

[6] Ibid., 95.

another being)—that Scotus excludes from animal seeds. An active potency of a different kind is what leads to the generation of something other than substantial forms. It is this kind of active potency that Scotus uses to explicate the Augustinian notion of "seminal reason": "In such a seed, there is some internal active power by means of which the seed is changed uniformly, as from a particular cause.... And this active quality, which the seed has from the progenitor, is the seminal reason."[7] Even if the effect of such potency is not the substantial form of the fetus, it is by no means inessential for the process of animal generation: The "seminal reason" is an "active quality induced into the seed, which is determinate and follows from a determinate form which is in the seed and transmutes it towards a determinate form."[8] In the sense that it changes the matter of the seed in such a way that the matter becomes capable of sustaining a substantial form, the effect of the active potency in the seed can be characterized as a "preliminary disposition" with respect to the animal that is generated.

Nevertheless, for Scotus, the form of the seed is an intermediary form in the sense that it ceases to exist once conception has taken place, and once the "seminal reason" is corrupted, only some accidental organization is preserved. This is why he answers the question of whether this active quality is active in substantial generations in the negative:

> Because as the substantial form of the seed decays, so does this active quality that follows the substantial form; hence, in the moment of corruption it does not generate, because then it does not exist, no more than the form of the seed from which it follows....[9]

In this sense, the effect of the active powers of a seed is limited: "[T]he active power in the seed is only capable of changing—not generating—substantial form; if something of the seed remains, it is less perfect than form."[10] This is why Scotus holds that, even if active potencies of the seed can be characterized as arising from the substantial form of the seed, the substantial form of the seed cannot be regarded as what generates the form of the living being. With respect to the living being to be generated, the seed is not a formal cause, but only a material cause.

The radical nature of Scotus's proposal can be measured by the fact that only very few early modern thinkers took up this line of thought. One notable exception is the Louvain-based physician Thomas Feyens, who is better known for his

[7] Ioannes Duns Scotus, *Opera omnia*, vol. 19 (Civitas Vaticana, IT: Typis Vaticanis, 1993), 159.
[8] Ibid.
[9] Ibid., 160–161.
[10] Ibid., 163.

contributions to the theory of monsters, and in particular for his claim that some malformations are due to maternal imagination.[11] However, while he invokes imagination for the acquisition of some traits, he does not take it to be responsible for the origin of the soul of the fetus.

Feyens considers several accounts of the origin of the soul of the fetus, all of which he deems to be unsatisfactory. He argues that it cannot be a power of an absent soul because powers cannot wander from one subject into another. He also argues that it cannot be a merely physical power. This is why:

> [I]f it were such a power, it would follow that the formation and change of the seed in the uterus were not vital actions: which is absurd.... The first premise is proved as follows: because from a non-vital power there cannot proceed a vital action. The second premise is proved as follows: because, what could be said more absurd than that the function of the seed in uterus is not an action of life? For the same power the change of the seed towards the nature of bone and flesh comes about as the one from which formation arises: for at the same time as the fetus begins to be formed, it begins to be nourished and to grow.[12]

Of course, a widely held view in medieval and early modern medical theory had it that the presence of vital processes from the moment of conception onward indicates the presence of souls in the seeds. Feyens rejects this view, and he does so in a way that goes significantly beyond Scotus, using a group of specifically medical arguments:

> Neither blood nor spirits are animated. Hence, nor is seed. The premise I assume on the basis of the common doctrine of all physicians.... The conclusion is proved as follows: Why would seed have to be more animated than those? They are produced in much more noble organs than the testicles; namely, in the brain, heart and liver; and if these are incapable of impressing a soul onto blood and the spirits, how could the testicles do this? Milk is produced in the third concoction in glandulous parts, in external parts, and is not animated: but the testicles are such parts: therefore, the seed that is generated in them can no more

[11] On Feyens's work in these areas, see L. J. Rather, "Thomas Fienus' (1567–1631) Dialectical Investigation of the Imagination as Cause and Cure of Bodily Disease," *Bulletin of the History of Medicine* 41 (1967): 349–367; Jan Papy, "The Attitude towards Aristotelian Biological Thought in the Louvain Medical Treatises during the Sixteenth and Early Seventeenth Century: The Case of Embryology," in *Aristotle's Animals in the Middle Ages and Renaissance*, eds. Carlos Steel, Guy Guldentops, and Pieter Beullens (Leuven, BE: Leuven University Press, 1999), 317–337.

[12] Feyens, *De formatrice foetus*, 92; see Gallego, *Opera*, 102.

> be animated than milk. Likewise, physicians deny a soul to spirits and blood because they are fluid bodies, not cohering with a whole, in no way organic, having no determinate magnitude and boundary of their substance. But seed is a body of this kind....[13]

Thus, it is the analogy between semen and other bodily fluids that are taken to be inanimate in the medical tradition that speaks against ascribing souls to seeds. Gallego's attitude toward these aspects of the thought of Feyens is clearly affirmative. But he takes Feyens's line of argument one step further by considering the reason why blood and milk are not animated. What matters, in his view, is the distinction between concoction and nutrition:

> The task of concoction in the precise sense is not to transmute ailment into the substance of what is nourished. For this latter task is the task of the power of nutrition; because the task of concoction is to assist and to prepare ailment for the nutritive power itself; and the task of the nutritive power is to animate it through and through.[14]

As Gallego explains, "the concocting power provides the assimilation of all parts of the ailment by educing from their potency some form that is intermediary between the vital form of this part, and the form that existed before in the parts of the ailment; this intermediary form is...the ultimate material disposition towards the vital form."[15]

Obviously, there is strong agreement among Scotus, Feyens, and Gallego with respect to the conception of animal seeds as merely material causes of generation. In spite of his agreement with the analysis of animal seeds as merely material causes of generation, Gallego is aware that neither Scotus nor Feyens is very strong on the issue of the origin of the substantial forms that they ascribe to living beings. In Scotus's view, inanimate forms are induced by the form of the heaven. But since he takes the form of the heaven to be inanimate, he does not invoke this explanatory pattern for the origin of the forms of living beings. Rather, in his view, vital forms are induced into the mixture of the seed and menstrual blood immediately by divine agency.[16] Likewise, with respect to human souls, Feyens accepts the doctrine that they are induced by divine agency. But Feyens does not come forth with any explanation of the origin of the souls of nonhuman animals. Gallego is not happy about this explanatory gap because

[13] Feyens, *De formatrice foetus*, 61–62.

[14] Gallego, *Opera*, 70.

[15] Ibid., 71.

[16] Scotus, *Opera omnia*, vol. 19, 163–164; see Gallego, *Opera*, 125.

he demands that the origin of the substantial forms of animals be explained through purely natural powers.

With respect to Feyens, he believes that the explanatory gap arises from an inadequate analysis of the role of material dispositions for the reception of substantial forms. Feyens's thesis is that "organization is not a disposition prior to the introduction of the soul."[17] This is why: "Being organic is nothing but being figured in this and that way; and organization is nothing but such and such a mode of figure in a subject and its parts: but figure is not a necessary disposition for the reception of natural forms...."[18] Understanding material dispositions in this way lends plausibility to the claim that there are various forms that are indeed independent of a particular disposition of matter:

> The forms of elements, forms of stones and metals are introduced into matter without any form. The real accidental forms such as heat, cold, light, rarity, density, softness, hardness, heaviness, lightness, color, taste, smell, and intentional forms such as the species of color, sound, smells, etc. all are received in the subject without that it be required that in the subject there is found any determinate figure and that all are indifferent to all figures; hence also the soul does not need it for its reception.[19]

As Feyens suggests, the relations of the soul to matter should be understood in analogy to the relation of these forms to matter. With respect to determinate figures of bodies of living beings, Feyens confirms this insight through a further consideration: "The figure of the body of living beings is the effect of the forms that exist in them; hence it is not a disposition prior to the forms . . . for a disposition is prior to the form, an effect posterior to it; but nothing can be prior and posterior at the same time."[20]

In contrast, Gallego takes the theory of animal seeds as merely material causes to imply that the seed has only received some material dispositions for the reception of the soul and the function of the parts of the fetus.[21] Moreover, in his view, an adequate solution to the problem of the role of material dispositions in animal generation will lead to an answer concerning the problem of the origin of animal souls. In his view, Suárez's theory of matter and form as incomplete entities is exactly what can provide solutions to both problems.

[17] Feyens, *De formatrice foetus*, 135.
[18] Ibid., 141.
[19] Ibid., 142.
[20] Ibid., 143.
[21] Gallego, *Opera*, 98.

3. Material Causes and Incomplete Entities

Suárez maintains that material dispositions cooperate actively in the production of substantial forms. This is his argument:

> An accident that is proportional to a substantial form can by its own nature be a suitable instrument for this task; for, even if with respect to its entity an accidental form is inferior compared with a substantial form, with respect to its way of being it has a proportion to it, if both depend on matter: and likewise they are commensurable with respect to disposition and form, and for the same reason they can be regarded as proportional with respect to instrumental power and action, as well as end. For, as it belongs to the nature of a substance to operate through accidents that are proportional to it, it belongs to the nature of a substance to come into being through accidental dispositions that are proportional to it. . . .[22]

Suárez calls a form of the type that he has in mind here a "material form"; and what is special about such a form is that it "depends in its being on matter."[23] But matter, too, in a certain sense depends on material form, and it is the concept of mutual dependence between matter and form that lies at the heart of Suárez's conception of matter and form as incomplete entities:

> [The union of material form and matter] insofar as it proceeds from form, is a medium or relation through which due to the form matter is actualized and a composite is put together; and in this way it is said to be the causality of form; but insofar as through this union the form adheres to matter and is sustained by it, it is a dependency of this form on matter. For there is such an intrinsic connection between such a form and union that they mutually depend on each other in different respects. [24]

What are the "different respects" that Suárez has in mind here? As to the dependence of form on matter, Suárez accepts the Scholastic view that "form necessarily requires a disposition of matter for it to be able to confer its formal effect."[25]

[22] Franciso Suárez, *Disputationes metaphysicae*, vol. 1 (Salamanca, ES: Ioannes & Andreas Renaut, 1597), 18.2.20 (cited according to disputation, section, and subsection; available online at googlebooks.com).

[23] Suárez, *Disputationes metaphysicae*, 15.6.10.

[24] Ibid.

[25] Ibid., 15.6.5.

But he maintains that the question of what the required disposition consists in can be answered only through a treatment of the relation between prime matter and quantity.[26]

In a recent work of commentary, Suárez's views on prime matter have been portrayed as being in great agreement with Aquinas's. It has been suggested that, for Aquinas and Suárez alike, prime matter (1) does not possess any active properties; (2) does not possess quantity; and (3) does not possess any qualities following from quantity.[27] Such an interpretation is highly inaccurate. It overlooks the fact that Suárez takes up Averroes's view that prime matter itself is capable of bringing forth extension, even if by itself it is incapable of defining spatial boundaries between individual things.[28] In fact, in Suárez's view, there are two possible ways in which prime matter could be thought of as bringing forth quantity: Either quantity is a necessary affection of a purely passive potency;[29] or prime matter—even if it does not act on other beings—can have an intrinsic capacity of bringing forth qualities by means of emanative causation.[30] Moreover, if prime matter is capable of bringing forth quantity, it is capable of bringing forth qualities that follow from quantity. This is suggested by the following passage:

> It is certain that simple substances that persist by themselves suffice to receive or materially cause accidents that are proportional to themselves.... When we say that these substances are by themselves sufficient for this, we do not exclude that they can materially cause one accident by means of another....

Here, Suárez distinguished between different levels of material accidents. Moreover, all accidents that are caused by first-level accidents are understood as being brought forth by the substance that brings forth first-level accidents. Thus, if prime matter is able to bring forth quantity, all accidents caused by quantity can be said to be brought forth by prime matter. In this sense, prime matter can produce material accidents.

In contrast, Suárez argues that material accidents cannot be caused by material forms alone. This flatly goes against the Thomistic view that ascribes both quantity and the qualities following from it to the union of matter with form. Suárez rejects this view because he believes that the substantial forms capable

[26] Ibid.

[27] John D. Kronen, Sandra Menssen, and Thomas D. Sullivan, "The Problem of the Continuant: Aquinas and Suárez on Prime Matter and Substantial Generation," *The Review of Metaphysics* 53 (2000): 863–885, 873.

[28] Suárez, *Disputationes metaphysicae*, 14.3.10; Suárez refers to Averroes's *Physics Commentary* I, 64.

[29] Suárez, *Disputationes metaphysicae*, 14.3.57.

[30] Ibid.

of informing matter lack subsistence on their own: "It is also certain that substantial forms that cannot subsist naturally are unsuitable to cause materially accidents by themselves alone: because what does not subsist in itself cannot support something else...."[31] This is the basic idea of Suárez's conception of substantial forms as incomplete entities—they are incomplete in the sense that they could not exist independently of some portion of prime matter that they inform. Suárez's argument for why substantial forms alone are incapable of bringing forth material accidents, however, is vulnerable to the objection that prime matter, no less than substantial forms, is incapable of subsisting by itself. In fact, Suárez acknowledges the dependence of matter on form, and there is a sense in which he regards prime matter, too, as an incomplete substance. Importantly, however, he distinguishes the structure of the dependence of matter on form from the structure of the dependence of form on matter:

> [M]atter has partial subsistence, hence it has some entity such that it can sustain some accidents... the dependence of matter on form is not the one true and proper cause of this piece of matter but only a cause that is naturally necessary for the preservation of matter; this kind of dependence, however, which is extrinsic to the entity of matter... does not prevent the entity of matter from being sufficient for sustaining accidents in the category of material cause.[32]

Thus, the dependence of matter on form does not exclude a sense in which matter possesses a kind of subsistence sufficient to sustain quantity. Still, material accidents not only depend on the subsistence of prime matter; they also, in a different sense, depend on the activity of substantial form: "Although quantity can be said to depend in a mediate way on the soul, insofar as matter depends on the soul, nevertheless this dependence is sufficiently replaced by the subsequent form...."[33] Most plausibly, what Suárez has in mind here is the Averroistic insight that determinate quantity depends on the presence of a substantial form. This conjecture is confirmed by his claim that matter cannot receive quantity "unless it is determined by form."[34] At the same time, the partial subsistence of prime matter implies that a determinate quantity does not depend on the presence of a particular substantial form. Any substantial form of the right type would do. In this sense, one substantial form can be replaced by another substantial form of the right type.

[31] Ibid., 14.3.3.
[32] Ibid., 14.3.12.
[33] Ibid., 14.3.19.
[34] Ibid., 14.3.42.

The upshot of Suárez's view is that there is a "mutual dependence and priority" of quantity and form:

> [Q]uantity and form can depend on each other in different material respects. For even if form with respect to matter is a formal cause, with respect to quantity, it is reduced to a material form... because it behaves like a disposition of matter such that matter can sustain quantity. Nor is it unsuitable that dispositions relating to different respects should have such a mutual dependence among each other, both because they are directed towards different formal effects that, under different headings, are necessary for the subject, as well as especially because they do not have this mutual order among each other immediately but with respect to the subject for whose being they are necessary. [35]

Gallego agrees with Suárez's general insight into the mutual dependence between quantity and form. Still, he believes that Suárez has not succeeded in fully explaining why matter cannot naturally receive a form unless it is extended. In his view, this is so because Suárez did not provide a fully satisfactory analysis of the notion of formal causation. As Suárez suggests, the causality of form is nothing other than the "actual union of form to matter" such that "the composite is the ultimate effect of this causality."[36] But while his view of formal causality makes clear why form is incomplete in the sense of being unable to exert formal causality without primary matter, it falls short of explaining why formal causality requires primary matter that is already extended due to its own (passive or active) potencies. The idea of formal causation as union of form and matter seems perfectly compatible with the idea that extension is produced, not presupposed by such a union. Moreover, the nature of this union remains puzzling. Clearly, only being where a portion of matter cannot be enough for union because then no composite substance would arise.

Gallego tries to remedy the shortcomings in Suárez's argument by using an alternative analysis of formal causation, an analysis that puts the nature of the activity of substantial forms at center stage. His basic insight is that it does not make sense to talk of a union of form and matter unless form acts upon a portion of matter. Gallego finds support for such a suggestion in Aristotle's *De caelo* II, 8, where Aristotle points out that heavenly bodies lack organs of animal motion but have the adequate shape for the motion specific to them (290a30–35). Gallego takes this passage to imply that "no created being has being for its own sake, that is, has being only for the sake of being, but it has being for the sake

[35] Ibid., 14.3.42.

[36] Ibid., 15.6.7.

of operation; and the operation... is more powerful and more useful than mere being, which is why it happens that the being of no created being can exist even for an instant without the action of organs."[37] Correspondingly, Gallego objects that "no substantial form unites with a body only for the sake of union but for the sake of carrying out some of its actions; hence, the soul unites with the body not for the sake of its first actuality, which is the substantial union, but rather for the sake of its most important and ultimate end."[38] And, as he emphasizes, in the case of animal souls, this ultimate end consists in organic actions.[39]

Thus, for Gallego, the production of substantial form requires two factors: (1) a specifically organic disposition of matter that is the outcome of the quantitative characteristics of prime matter, and (2) an external agent that produces a natural form-matter composite. We look into Gallego's views on the nature of this external agent and its mode of operation in the following section. For now, it is important to get a grasp on the characterization that Gallego gives of the substantial forms generated in this way. With respect to the substantial form that animates seminal matter at an initial stage, Gallego is clear that it must be a kind of vegetative soul.[40] He distinguishes such souls from both the merely natural and the entirely spiritual forms, but he is never very explicit as to the nature of vegetative souls. However, he notes that Galen identifies the potency capable of bringing forth change in an already formed organism with the temperament of primary qualities.[41] As Gallego suggests, the formative faculty of the fetus, too, could be understood as something that "follows" the temperament of primary qualities.[42] Moreover, he is explicit that because of their relation to body parts, animal souls are themselves quantitative entities:

> [I]n order to be able to inform some body... it is not enough that the soul be simply a quantum, but it is necessary that it actually has the same extension of quantity as is necessary for the secondary actuality of the form.[43]

This conception has far-reaching consequences involving the mereological structure of animal souls. Contrary to the widely accepted doctrine that the soul is whole in the whole body and whole in each of the body parts, Gallego holds that the relation of animal souls to animal bodies is more complicated.

[37] Gallego, *De naturali animarum origine*, 42–43.
[38] Gallego, *Opera*, 142.
[39] Ibid.
[40] Ibid.
[41] Ibid., 119; Gallego's reference is to book I of Galen's *De naturalibus facultatibus*.
[42] Gallego, *Opera*, 174.
[43] Ibid., 142.

There is, in fact, a sense in which he believes that the soul is related to the whole body: "[E]ach soul requires a whole that is indivisibly prepared in the way described, as the adequate being that can be perfected, because it is proportional to the whole and not the single parts. . . ."[44] The relevant sense in which the body is an indivisible whole can be gleaned from Gallego's acceptance of Galen's notion of instrumental cause: According to this notion, single body parts fulfill specific functions, but they do so in the mode of "instruments," that is, their agency depends on the agency of other body parts. An organic body forms a whole in the sense that the proper operation of one body part depends on the proper functioning of some other body parts, and vice versa.[45] In this sense, an animal soul requires an organic body that is a genuine whole. At the same time, specific bodily dispositions correspond to specific functions of the soul:

> [T]he whole soul is educed from the whole body, but in such a way that each of its parts is educed from the part of the body to which it is said to have an aptitude, due to the suitable disposition of this body part. For as the whole material soul is said to have a relation to the whole body through the variety of dispositions of a shared nature, so is each part of the form said to have a relation to one particular part of the whole body, due to a singular disposition and a partial distinction from other parts of the same body. . . .[46]

This is why animal souls "are accidentally quantitatively extended according to the parts of the whole, in which they are not simply but in a diminished, secondary way, or in a certain respect; and so it comes about that souls exist not as perfectly in the part as in the whole. . . ."[47] Moreover, that an animal soul "depends on matter in its material being"[48] implies that it is divisible through the division of bodies:

> [A]ll souls that are generated by a corporeal agent and that have such a dependence on matter that they perish together with it and are measured through its quantity, are divisible through the division of quantity: for whatever is material with respect to its origin is disposed and affected according to the affections of the receiving body.[49]

[44] Gallego, *De naturali animarum origine*, 32.

[45] Gallego, *Opera*, 23; Gallego's references are to Galen, *De elementis* I, 9, and *De methodo medendi* I, 6.

[46] Gallego, *Opera*, 143.

[47] Gallego, *De naturali animarum origine*, 32.

[48] Ibid., 10.

[49] Ibid., 35.

The conception of animal souls as extended and divisible beings is fully in line with Gallego's suggestion that the formative principle internal to the fetus "follows" the temperament of primary qualities. Likewise, it ascribes to animal souls an ontological status that is compatible with the idea that they are "educed" from the potencies of matter. With this characterization of animal souls in hand, we now can look into the characterization of the required external agent and the physiological processes that Gallego regards as responsible for their formation.

4. Incomplete Entities and the Formative Power of the Uterus

Again, Gallego takes up a central insight from Suárez, who ascribes to material accidents an active role in the production of substantial forms but does not believe that material accidents alone could bring forth substantial forms. As Suárez puts it:

> And even if there is a natural sequence between this disposition and the form, it is necessary that this sequence happens due to some other efficient cause. For this sequence is not the emanation of the form from the disposition...but it is only the necessary and immediate succession following such a disposition, a succession that necessarily must be founded in the efficacy of another cause, which produces one immediately after the other.[50]

Gallego accepts that for the generation of substantial forms in addition to the disposition of matter, there has to be the agency of an active principle external to the entity to be generated. In distinguishing between the active power capable of educing animal souls from matter and the form of the fetus, he closely follows Suárez's insight that "[a] form...that is educed from the potency of matter is not properly the formal cause of the action through which it is educed...because it is its end...."[51] But as Gallego notes, Suárez is not forthcoming in telling what the external cause of substantial forms in the case of animal generation is. At this juncture, Gallego takes up the Galenic conjecture that the uterus possesses a formative faculty that contributes to the formation of the fetus.

By the early 17th century, this conjecture was far from uncontested. For example, Feyens flatly denies that the generative power of the uterus can concur to generation. One of Feyens's arguments concerns the notion of formative

[50] Suárez, *Disputationes metaphysicae*, 14.3.30.

[51] Ibid., 15.6.10.

potency: "Formation is a vital action, that is, an act of life; but acts of life take place by means of an internal principle, not an external one; hence, not by means of the uterus, which is external to the seed...."[52] Another argument has to do with the observation that the formation of the fetus does not start with the outer parts of the fetus:

> The seed begins to be formed with respect to all of its parts equally, internal parts as quickly as external parts. Hence, the power that forms is within it. For, if it was not and came from the outside, the external parts would be formed more quickly than the internal parts; and those more quickly that are closer to the place where the power resides than those that are farther away from it.[53]

In response to these objections, Gallego accepts that the fetus possesses an internal formative power but denies that this power develops out of an internal formative power of the seeds. In his view, the formative power of the fetus is brought forth through the agency of the uterus, together with a suitable disposition of seminal matter. The upshot of this theory is that the vegetative soul of the fetus is not an entity formed by the uterus and then implanted into the seminal matter, but rather an entity that depends both on the agency of uterus and the material disposition of the seminal matter:

> [T]he whole being of material souls arises from the progenitor effectively and from matter subjectively; and this is why they are said to be educed from the potency of matter. Which means nothing other than that these souls do not come into being separately from matter, and that a whole by itself comes into being through the union of form and matter; for as the soul of a dog, for example, has a natural inclination towards matter disposed for it: so does the matter disposed by the dog have a natural inclination towards the dog-form. Such a soul, however, is said to be educed from the potency of matter because it is brought into actuality through transmutation brought forth by the agent.[54]

Still, a crucial question remains: What are the physiological processes by means of which the uterus educes a living form from the potency of seminal matter? Gallego characterizes what happens in the uterus as mere "addition or subtraction of matter."[55] Moreover, as far as the production of the ultimate disposition

[52] Feyens, *De formatrice foetus*, 23.

[53] Ibid.

[54] Gallego, *De naturali animarum origine*, 7.

[55] Ibid., 8.

of matter for the reception of form goes, Gallego characterizes the activity of the uterus in terms of qualitative change:

> [N]ot only nature but also art that imitates nature, as often as it wants to inform some piece of matter, always starts with introducing the necessary dispositions. For as often as fire tries to introduce its form into a piece of wood, we experience that it communicates first its qualities to this piece of wood, and finally introduces its form. . . .[56]

In this context, he mentions approvingly a passage from an Arabic Galenist, Hali Abbas ('Ali ben 'Abbas, d. c. 994), according to whom the formative power of the uterus should be characterized as the "force of changing and dividing the seed into the various members of the fetus with different temperaments and figures."[57] Gallego also follows Hali in ascribing an attractive power not only to animated seminal matter but also to bodily organs, including the uterus.[58] But while Hali invokes the attractive power of the uterus only to explain where conception takes place, Gallego extends this idea to explain how the uterus forms the fetus: "[I]n the body there is no organ that by its similarity and natural inclination has the power of attracting some useful substance only for the sake of attraction but rather for the sake of fruition; and because fruitions are varied, and in the natural living bodies they never happen without a change of the attracted thing through a power impressed on them by the attracting thing."[59] The process that takes place in the uterus thus should be seen in analogy with what happens in other organs. Animal generation is understood here as a kind of metabolism, and this applies especially to the generation of animal souls:

> The ultimate form follows naturally from the ultimate disposition out of the necessity of one and the same action. For the form does not follow from the dispositions themselves because these would be operative in the introduction of form, but rather because the power of what gives the ultimate disposition cannot give to this piece of matter, in addition the ultimate disposition, anything other than substantial form itself.[60]

This passage suggests that the action of the uterus that educes the vegetative soul is continuous with the action of the uterus that leads to the ultimate disposition

[56] Gallego, *Opera*, 139.

[57] Ibid., 148. The reference is to Hali Abbas, *Regalis dispositio* (Venice: Johannes de Nigro, 1492), fol. 25v (available online at googlebooks.com)

[58] Gallego, *Opera*, 148; see Hali, *Regalis dispositio*, fol. 25r.

[59] Gallego, *Opera*, 148.

[60] Ibid., 158.

of seminal matter. It is just one more step in the addition and subtraction of material parts—the step that brings forth the temperament upon which the vegetative soul follows. If this is what Gallego has in mind, the production of the ultimate disposition of seminal matter and the "eduction" of the vegetative soul are processes of one and the same kind.

Does Gallego connect this general theoretical framework with any determinate view as to the determination of sex and the acquisition of individual traits? He does, and he does so by departing from a view widely held in his time: the view that the seed is a homogeneous substance (i.e., a substance lacking any structural differences among its single parts). Gallego is aware that if seeds are to play a significant role as merely material causes of animal generation, they must contain parts of varying structure, even if their differences are not perceptible. The agency of the uterus, in his view, depends on the differences among the parts of seminal matter:

> In the [male] seed there can be diverse parts such that it together with the female seed and the menstrual blood collected together in the uterus can produce body parts of such diverse nature as well as several fetuses of the same or different sex. And if the matter is perfectly concocted and at the same time spirituous and copious, nature without doubt decides because of the amount and perfection of matter to form several male fetuses in all parts of the uterus; but if matter is copious and endowed with contrary quality, it starts from the beginning to form many female fetuses. However, if matter is copious and intermediary with respect to quality, nature knows in the first instance to separate the parts of matter, which are suitable for a male fetus, and to collect all of it in one place, and those that are suitable for a female fetus in another, and afterward to begin with the formation of both bodies.[61]

Of course, this passage is not free from the biological chauvinism that Gallego shares with many of his predecessors and contemporaries when he tags the differences between male and female seeds in terms of degrees of perfection. However, he regards his account as fully in line with the explanation of the determination of sex proposed by Galen, and there we encounter a reduction of talk of degrees of perfection to a conception of degrees of structural differences that is more value neutral: As Gallego notes, according to Galen "a seed that is less perfect in motion becomes a female, a seed more perfect a male. Yet, he reduces the more or less perfect to a greater or smaller heat, and says that all natural actions can be reduced to this physical principle."[62] Speaking of greater

[61] Ibid.
[62] Ibid.

or smaller heat—qualities that are internal to the seeds—suggests that what Galen has in mind here are motions internal to the seeds. Arguably, Galen here applies Aristotle's conjectures concerning the role of internal motions of seeds in inheritance from *De generatione animalium* IV, 3. But while Aristotle explicitly ascribes to internal motions an information-bearing character only in the context of his theory of the processes in sensory organs,[63] Gallego characterizes the role of material structures internal to the seeds as information-bearing in a quite specific sense:

> From all of the mentioned philosophical foundations it can be derived that the female acts in a mediate way through the uterus, as a univocal cause according to the disposition of matter, not only with respect to accidental similarity but also with respect to similarity of sex, insofar as it is moved towards the generation of several fetuses or a single fetus, guided only by the indication given by the quantity of matter. . . .[64]

Strikingly, material causes here are characterized with the help of a notion that belongs to the theory of signs, and especially to medical semiotics: the notion of indication. As Gallego characterizes them in another medical text, indications are bodily dispositions that fulfill two further criteria: (1) They are immediate causes of imminent diseases, and (2) they are the basis upon which the physician takes precautions such as administering medicaments.[65] This characterization is fully consistent with the Galenic notion of indication: Indications are derived from knowledge of the nature of diseases and their causes, and they allow the physician to use medicaments whose qualities are contrary to the qualities of the causes.[66] Of course, there are dissimilarities between medical indications and indications in the case of animal generation. Of course, in the case of indication in the context of animal generation, there is no agent that has intellect and cognition. Also, the signs that indicate something are not causally connected to anything that deviates from its proper nature. Accordingly, the agent in question does not fulfill the task of correcting anything. Nevertheless, the analogy between material structures internal to the seed and medical indications is

[63] See *De generatione animalium* V, 1. Devin Henry has argued that this could be sufficient grounds for ascribing an information-bearing character to the internal motions in seeds, as well; see Henry, "Aristotle on the Mechanism of Inheritance," *Journal of the History of Biology* 39 (2006): 425–455.

[64] Gallego, *Opera*, 158.

[65] Ibid., 303. The most detailed study of medical semiotics in Gallego's time is Ian Maclean, *Logic, Signs, and Nature in the Renaissance: The Case of Learned Medicine* (Cambridge, UK: Cambridge University Press, 2002).

[66] Gallego, *Opera*, 307.

illuminating because material structure in the seed can be understood in analogy to natural signs that signify according to the causal relations in which they stand: Medical indications are natural signs because they are causally related to an imminent disease. Analogously, material structures in seeds are natural signs because they are causally related to the traits and sexual characteristics of the parents. And as in the case of the physician interpreting medical indications, the uterus fulfills the role of an external agent that is guided by the information carried by a natural sign. Thus, only together with the agency of the uterus are the material structures internal to seeds capable of bringing forth animal souls. But this is exactly what Gallego's view of both animal souls and animal seeds as incomplete entities would have us expect.

5. Conclusion

Now it should be clear that Gallego had some quite innovative ideas about how to fill in the explanatory gaps left open by theories of animal seeds as merely material causes of biological reproduction, as suggested by Scotus and Feyens. Gallego has no qualms about the idea of divine concurrence in animal generation and other natural processes, as long as divine agency remains on the level of secondary causes. But he rejects the idea that God could be regarded as the primary cause of animal souls and offers an explanation of the origin of animal souls that invokes only natural primary causes. In his view, the theory of incomplete entities is what can explain how animal souls can be educed from the potency of matter and, as we have seen, thinking of seminal matter and animal souls as incomplete entities can be explicated in terms of the following intricate web of dependence relations:

(i) The dependence of determinate extension on the agency of form: Even if prime matter is capable of bringing forth extension, it is incapable of individuating particular boundaries of extension;
(ii) The dependence of the agency of form on extension: The union of form and matter presupposes extension (of the suitably structured kind) because the goal of this union consists in the specific activities of the form;
(iii) The dependence of both the ultimate preparation of matter and the final eduction of form from matter on an external agent; and
(iv) The dependence of the agency of an external agent on the structure of seminal matter.

Gallego identifies the external agent relevant for the generation of animal souls as the uterus and understands the process that leads to the generation of material forms in analogy with metabolism taking place in other bodily organs. In this way, as in Scotus and Feyens, animal seeds still function merely as material causes of reproduction. At the same time, Gallego goes beyond Scotus and Feyens by ascribing to such material causes enough internal structure to guide the development of the fetus in such a way that they account both for trait acquisition and the determination of sex.

8

Biology and Theology in Malebranche's Theory of Organic Generation

KAREN DETLEFSEN

1. Introduction

> The sketch by this philosopher [Descartes] can help us understand how the laws of motion suffice to cause the parts of animals to grow little by little. But that these laws are able to form them and connect them all together, is something no one will ever prove (OC XII, 264/DMR XI, 205).[1]

In this passage from his *Entretiens sur la métaphysique et sur la religion*, Nicholas Malebranche has seized upon one of the problems in natural history (and later, biology) that has proven, historically, to be among the most difficult—the question of the origin of organic forms. In E. S. Russell's evaluation, it is the problem of explaining the development of a living individual from something seemingly simple and uniform to a complex and organized creature. Further, the problem is to explain how a living creature is able to be unified at a time and identical through times despite its evident material diversity and mutability.[2]

[1] All references to Malebranche's works are to the following editions with their respective abbreviations: Nicolas Malebranche, *Dialogues on Metaphysics and on Religion* (reprint, Cambridge, UK: Cambridge University Press, 1997), trans. David Scott, ed. Nicholas Jolley, cited as DMR with dialogue section and page number; Malebranche, *Oeuvres complètes de Malebranche*, 21 vols. (Paris: J. Vrin, 1958–66), series director Henri Gouhier, various editors, cited as OC with volume and page number; Malebranche, *The Search After Truth* (Columbus, OH: Ohio State University Press, 1980), trans. Thomas M. Lennon and Paul J. Olscamp, cited as ST with page number.

[2] E. S. Russell, *The Interpretation of Development and Heredity* (Oxford, UK: Clarendon Press, 1930), 1.

Malebranche 'solves' the problem by denying it altogether. Generation of new lives does not occur. Each animal, plant, and human body is preformed, fully created by God at the Creation as a miniature of the eventual adult it will become, awaiting only its appropriate moment to grow into its adult size.[3] All these preformed organisms are stored from the Creation in the ovaries of the first female of each species (ovism).[4] The process of growth or augmentation of these organisms, a sort of swelling, is entirely possible by mechanical means. What we observe as and even call generation is simply the becoming visible of what was previously invisible, but it is not a new becoming, a new formation, and there is no development in any meaningful sense of the idea of development. That is, there is no coming-into-being of the completed organic form itself—that complex, organized, and unified living being. Nor is there a problem of accounting for the simplicity or identity of a creature; it was created in order to be intrinsically unified. This theory, termed "preexistence" by some,[5] was a prevalent theory of organic generation in natural philosophy for about a century, from the 1670s when Malebranche first clearly sketched it in its early modern form.[6] Preexistence is often contrasted with epigenesis, the theory that

[3] On Malebranche's theory of preexistence, see, e.g., Karen Detlefsen, "Supernaturalism, Occasionalism, and Preformation in Malebranche," *Perspectives on Science* 11.4 (2003): 443–483; Andrew Pyle, *Malebranche* (London: Routledge, 2003), chapter 7; Andrew Pyle, "Malebranche on Animal Generation: Preexistence and the Microscope," in *The Problem of Animal Generation in Early Modern Philosophy*, ed. Justin Smith (Cambridge, UK: Cambridge University Press, 2006), 194–214; André Robinet, *Malebranche de l'académie des sciences* (Paris: Vrin, 1970), section VI, chapter 2; and Paul Schrecker, "Malebranche et le préformisme biologique," *Revue internationale de Philosophie* 1.1 (1947): 77–97.

[4] Ovism is proposed by Swammerdam, Régnier de Graaf, John Ray, Pierre-Sylvain Régis, Bernard Le Boyer de Fontanelle, and Albrecht von Haller in addition to Malebranche. The version of preexistence that maintains that the organisms are housed in the first male of each species is termed "animalculism" or "spermism." A third version of preexistence, panspermism, maintains that organisms are scattered throughout nature. Animaculism's most famous adherents include Leeuwenhoek, Hartsoeker, and Leibniz (usually). Panspermism is embraced by Claude Perrault and La Mettrie.

[5] For different terms used for the theory I call "preexistence"—most notably "preformation"—see Jacques Roger, *The Life Sciences in Eighteenth-Century French Thought*, trans. Robert Ellrich, ed. Keith R. Benson (Stanford, CA: Stanford University Press, [1963] 1997), 259–260; and Peter J. Bowler, "Preformation and Preexistence in the Seventeenth Century: a Brief Analysis," *Journal of the History of Biology* 4.2 (1971): 221–222.

[6] See, e.g., OC XII, 253-4/DMR XI, 196. There is some dispute as to whether Malebranche or Swammerdam (or even Malpighi) is to be credited with the revival of the theory in its 17th-century form. See Howard B. Adelmann, *Marcello Malpighi and the Evolution of Embryology*, 5 vols. (Ithaca, NY: Cornell University Press, 1966), vol. II, 869–870; Richard Aulie, "Caspar Friedrich Wolff and His 'Theoria Generationis,' 1759," *Journal of the History of Medicine* 16 (1961): 124–144; Bowler, "Preformation and Preexistence," 234, fn. 34, and 237; Peter J. Bowler, "The Changing Meaning of "Evolution," *Journal of the History of Ideas* 36.1 (1975): 95–114; Daniel C. Fouke, "Mechanical and 'Organical' Models in Seventeenth-Century Explanations of Biological Reproduction," *Science in Context* 3.2 (1989): 365–381; and Edward Ruestow, "Piety and the Defense of Natural

maintains that each new organism is formed anew, the parts emerging successively, and often due to a *self*-generative capacity.[7] Many believe that epigenesis finally triumphed near the end of the 18th century, putting an end to the reign (or tyranny) of preexistence.

Why would so many smart people in the early modern period embrace such a seemingly bizarre theory of generation? In this chapter, I answer that question by examining three general categories of reasons why early modern natural philosophers endorsed preexistence, categories I will call theological, scientific,[8] and non-theological metaphysical (for ease of expression, I will simply call this last "metaphysical"). Thinkers who endorse preexistence often have multiple reasons from more than one of the above categories for their endorsement of the theory. And crucially, these categories are not always neatly distinct from one another—a point that will become amply clear in the second part of this chapter. Nonetheless, these three categories of reasons encompass most of the arguments given by those in the early modern period for their embrace of preexistence. In the second part of the chapter, I turn to an examination of Malebranche, who holds all these motivations save one (the metaphysical) in his own embrace of preexistence. That is, Malebranche maintains preexistence for theological *and* scientific reasons, and his scientific reasons are both that he is committed to mechanism and that he acknowledges the biological nature of the organic being. Indeed, I argue that the best characterization we can give of Malebranche's argumentative strategy is that he draws on all three of these strategies in an "all things considered" approach to provide strong evidence (but not decisive arguments) for preexistence as the most likely theory of generation.

Order: Swammerdam on Generation," in *Religion, Science and Worldview*, eds. Margaret Osler and Paul Farber (Cambridge, UK: Cambridge University Press, 1985), 231ff. I say "revival" of the preexistence theory in the early modern period because it was in existence in medieval philosophers such as Augustine and Bonaventure as well.

[7] William Harvey coined the term "epigenesis" and developed that theory in the 17th century (see, e.g., William Harvey, *Disputations Touching the Generation of Animals*, trans. Gweneth Whitteridge (Oxford, UK: Blackwell Scientific Publications [1651] 1981), 204. Some commentators believe that epigenesis as a biological theory can be traced back to Aristotle (e.g., A. L. Peck, Introduction to *Generation of Animals*, by Aristotle, trans. A. L. Peck (Cambridge, MA: Harvard University Press, 1963); and Anthony Preus, "Science and Philosophy in Aristotle's *Generation of Animals*," *Journal of the History of Biology* 3.1 (1970): 1–52. Although it is true that Aristotle believes that foetuses generate by a sequential formation of parts, there are crucial theoretical differences between his and Harvey's theories, which make it problematic to conflate them.

[8] In using the term "scientific" for one of the categories of reasons, I am mindful that this is a modern term unfamiliar to those of the 17th and 18th centuries. Moreover, the closest 17th-century discipline to what we call "science" (i.e., natural philosophy) embodies not only modern-day science, but metaphysics and theology as well. For this reason, I emphasize that the three categories of reasons given for preexistence are not neatly distinct. Yet there are arguments for preexistence that emphasize natural explanations for phenomena above all else, and these arguments are those I will call scientific.

2. Why Preexistence? Three Categories of Reasons

In this section, I examine three broad forms of arguments in favor of preexistence, arguments found in the writings of those in the 17th and 18th centuries who embrace the doctrine of preexistence. I also underscore that, though one form may dominate in any given argument, these three general forms of argument are not neatly distinct, and this blending of the theological, metaphysical, and scientific is to be expected in the age of natural philosophy (as distinct from science).

2.1. Theological Arguments

There are two primary theological defenses of preexistence that I examine. The first is the original sin defense, and the second is essentially a version of the teleological argument for God's existence.

According to the original sin defense of preexistence, the thinker acknowledges the Biblical idea that we are all stained by original sin, and it was to remove this sin that Jesus Christ sacrificed his life for us. But one may wonder *how* we all came to be stained by original sin when we were not there at the moment of sin to be held responsible for it. Preexistence, however, holds that we *were* all present at the beginning to be stained by sin—indeed, we were all housed in the ovaries of Eve or the "loins" of Adam, the very individuals who committed that sin. And so, according to this argument, preexistence finds support from the fact that it can help explain the theological belief that we are all implicated in that free choice and action of Eve. As Swammerdam writes: "[E]ven original sin may stand on this principle [of seeds being present in the father's loins], since all mankind have been laid up originally in the loins of their first parents."[9]

The second theological argument I examine here is a version of the teleological argument for God's existence. Heeding the intricate and beautiful design of living bodies that are, moreover, organized to fulfill manifest purposes (e.g., eyes fulfill the purpose of seeing), a theologically minded early modern philosopher can point to these natural facts as the most obvious evidence to be found in the natural world for the direct role of a purposeful, powerful, and intelligent designer—that is, God—in the creation of these bodies. Many in the 17th century did argue in this way.[10] It is, of course, a further step to say that God created

[9] Jan Swammerdam, *Book of Nature, or the Natural History of Insects* (London: n.p., 1738).

[10] As just three examples, see Robert Boyle, "A Disquisition about the Final Causes of Natural Things," in *The Works of Robert Boyle*, vol. 11, eds. Michael Hunter and Edwards B. Davis (London: Pickering and Chatto, [1688] 2000), 79–151; Pierre Gassendi, "Fifth Set of Objections to Descartes' *Meditations in First Philosophy*," in *The philosophical Writings of Descartes*, vol. II, trans. John Cottingham, Robert Stoothoff, and Dugald Murdoch (Cambridge, UK: Cambridge University Press, [1641] 1984), 215; and Malebranche (e.g., OC XII, 247/DMR X, 189).

all organisms, fully formed, at the Creation, rather than throughout the duration of the natural world. But one possible way to this stronger conclusion is to claim that preexistence leaves a substantial role for God in the process of generation without thereby calling upon the deus ex machina each and every time a new organism is formed. This second theological argument is closely tied with the first of two scientific arguments I focus on, and I turn to these now.

2.2. Scientific Arguments

There are two primary scientific arguments for preexistence that I examine. The first is the argument from the limits of mechanism taken as a form of explanation of natural phenomena, and the second is the argument from biology, an argument that acknowledges the unique character of living beings.

In the early half of the 20th century, historians of science were not kind to the preexistence doctrine. For example, F. J. Cole, though generous in space in his discussion of preexistence is not generous in judgment, concluding that the "contributions of the early naturalists to the doctrine of generation [are] ... too brief.... That this meagre record is the direct consequence of the paralysing influence of the Preformation [preexistence] Doctrine is beyond question.... A speculation which makes a picturesque or academic appeal to the imagination ... is assured of the warmest reception. The Preformation Doctrine is an excellent illustration of this point."[11] Preexistence was rescued from ignominy in the 1960s by a series of historical studies that demonstrated its roots in the metaphysics and natural history of the early modern period. This was part of a general transition to more sympathetic, contextualized approaches to the history of science. Paramount among these studies is Jacques Roger's monumental study, *Les Sciences de la vie dans la pensée française du XVIIIe siècle: la génération des animaux de Descartes à l'Encyclopédie*, recently translated into English under the title *The Life Sciences in Eighteenth-Century French Thought.*[12]

[11] F. J. Cole, *Early Theories of Sexual Generation* (Oxford, UK: Clarendon Press, 1930), 201–202. Cf. Russell, *The Interpretation of Development*, 29ff.; Arthur William Meyer, *The Rise of Embryology* (Stanford, CA: Stanford University Press, 1939), 84ff.; and Joseph Needham, *A History of Embryology*, 2nd ed. (New York: Abelard-Schuman, 1959), 223ff.

[12] Roger, *The Life Sciences*. Other prominent studies in this early wave that dealt seriously with preexistence include books by Adelmann, *Marcello Malpighi*; Elizabeth Gasking, *Investigations into Generation, 1651–1828* (Baltimore, MD: Johns Hopkins University Press, 1967); Jane Oppenheimer, *Essays in the History of Embryology and Biology* (Cambridge, MA: MIT Press, 1967); and François Jacob, *The Logic of Life: A History of Heredity*, trans. Betty E. Spillmann (Princeton, NJ: Princeton University Press, [1970] 1993); and articles by Charles W. Bodemer, "Embryological Thought in Seventeenth-Century England," in *Medical Investigations in Seventeenth-Century England* (Los Angeles: William Andrews Clark Memorial Library, 1968); Bowler "Preformation and Pre-existence"; François Duchesneau, "Haller et les théories de Buffon et C.F. Wolff sur l'épigenèse,"

In his text, Roger builds the case that, faced with the limitations of the nascent and unsophisticated brand of mechanism to which many 17th- and 18th-century philosophers and natural historians were committed, preexistence was the most sensible answer to the problem of generation—the apparent new creation of a living being from parents of the same kind. For my purposes, I define mechanism as the belief that all changes at the phenomenal level—that is, all changes we experience—are due to the lawful motion and contact of sub-visible matter that is inherently inert and quantitatively, not qualitatively, defined. According to Roger's thesis, bits of matter this austerely defined, together with a few very simple laws of motion, could never give rise to phenomena as complex as the generation of living beings. The evidence for mechanism's inability to account for generation was produced by Descartes when he attempted to provide such an explanation and, many believed, failed abjectly. Descartes's explanation for the phenomenon of organic generation runs roughly as follows: The seminal fluid from the female and male mix in the uterus. They begin a rapid vortical motion due to extreme heat caused by a sort of fermentation upon the mixing, and from this mechanically moving liquid, the heart, then brain, and then other body parts emerge one by one until a living being is formed (e.g., AT XI, 253ff./ CSM I, 322ff.).[13]

Many natural philosophers who came after Descartes found this explanation entirely improbable. It is unlikely that a complex and well-organized being could develop this way just once, let alone with the frequency and respect for natural kinds that we regularly observe. The primary problem is one of chance. Should a living being with its incredible complexity be brought into existence through the movement of unorganized and homogenous matter according to a few simple laws, this would be a matter of sheer chance, here defined in contrast with purpose.[14] But the process of foetal formation happens all the time, quite regularly and in a highly deterministic fashion. One cannot realistically explain such a

History and Philosophy of the Life Sciences 1.1 (1979): 65–100; Shirley Roe, "The Development of Albrecht von Haller's Views on Embryology," *Journal of the History of Biology* 8.2 (1975): 167–190; and Shirley Roe, "Rationalism and Embryology: Caspar Friedrich Wolff's Theory of Epigenesis," *Journal of the History of Biology* 12.1 (1979): 1–43.

[13] All references to Descartes's works are to the following editions with their respective abbreviations: René Descartes, *Oeuvres de Descartes,* 11 vols., eds. C. Adam and P. Tannery (Paris: J. Vrin, 1964–1976), cited by AT followed by volume and page number; and Descartes, *The Philosophical Writings of Descartes,* 2 vols, trans. John Cottingham, Robert Stoothoff, and Dugald Murdoch (Cambridge, UK: Cambridge University Press, 1985), cited by CSM followed by volume and page number.

[14] Chance, here, is not contrasted with randomness. Descartes's account is fully deterministic, and therefore not random. But given the unorganized nature of the matter of the universe upon creation (according to Descartes's "chaos fable"), and given the indiscriminate nature of the laws of motion, then there is seemingly no purpose reflected in Descartes's theory of generation.

process in terms of something as chance-laden as initially unorganized matter in lawful motion. Generation, probably more than any other natural phenomenon, threatens early modern mechanism, here taken as a scientific theory (i.e., theory meant to explain natural phenomena). And so, in lieu of abandoning strict mechanism by positing some active and purposeful or intelligent agent in nature, many natural philosophers adopted the belief that God—and not something in the natural world—is responsible for the creation of these forms.[15] Stephen Jay Gould picks up this theme when explaining his own abandonment of the "silly student's view" that preexistence was a "nonsensical doctrine": "I came to understand that the leading preformationists had been, ironically, defenders of the general mechanistic attitude that modern science now honors, whereas the epigeneticists had tended to favor the vitalistic views now anathematized."[16]

The connection between science and theology is especially clear when thinking about this first scientific argument. The issue of chance—in contrast to *purpose*—just discussed establishes this, for purpose and a mind (God's) that must be present for there to be purpose, is at the heart of the second (teleological) theological argument previously mentioned. In fact, Catherine Wilson has noted (in conversation) that the supposed inability of mechanism to explain generation is not likely the primary reason at all for the prevalence of the preexistence doctrine in the early modern period. She believes that theological motivations—and not issues with mechanism—are at the core of the popularity of preexistence. She notes that some early moderns opposed to the mechanical philosophy for theological or moral reasons (e.g., More, Cudworth, and Kant) use the especially difficult problem of generation to undermine the mechanical philosophy generally. Others have noted that mechanism essentially secures a crucial role for God in natural philosophy—thus necessarily linking mechanism and theology—because a natural world comprised exclusively of passive matter must include God in order to explain basic phenomenological facts about that world, such as motion.[17] So the emergence of preexistence occurred because this theory was the only one that completely secured a role for God in 'natural' philosophy.

Nonetheless, I take the second theological (from teleology) argument and this first scientific (from mechanism) argument to be two distinct arguments, however closely related. The teleological argument for preexistence takes the goal to be the proof of God's existence, whereas the argument from mechanism takes the goal to be an account of generation (a natural phenomenon) from

[15] E.g., Roger, *The Life Sciences*, 128–129.

[16] Stephen Jay Gould, foreword to *The Ovary of Eve: Egg, Sperm and Preformation*, by Clara Pinto-Correia (Chicago: University of Chicago Press, 1997), xiv–xv.

[17] See Keith Hutchison, "Supernaturalism and the Mechanical Philosophy," *History of Science* 21 (1983): 297–333.

within the confines of mechanism (taken as an explanatory account of natural phenomena). For this reason, I treat them as two distinct kinds of arguments.

I will call the second scientific argument the "biological argument," and it runs as follows: Recognizing the unique kind of structural unity characteristic of living beings, and attributing that unity to the capacity of living beings to function to maintain that structure (i.e., to function toward the end of self-preservation), early moderns often located the origin of these functional capacities in God; God created living beings precisely so that they could function for the end of self-preservation. We find an example of this form of scientific argument in Charles Bonnet:

> One needs no Morgagni, no Haller, no Albinus [experimentalists who believed they had provided empirical proof for preexistence] to see that all constituent parts of the body are so directly, so variously, so manifoldly, intertwined as regards their function, that their relationship is so tight and so indivisible, that they must have originated all together at one and the same time. The artery implies the vein, their operation implies the nerves, which in their turn imply the brain and that by consequence the heart, and every single condition a whole row of other conditions.[18]

The further belief that *God* must create organisms in order to explain their biological functions is motivated by the following: First, these functional capacities are taken to be truly teleological, and second, among the teleological options available to the early modern thinker, those that would be acceptable to a devotee of mechanism posit God as the intelligent source of the purposeful functions of organisms. Further, as suggested by the passage from Bonnet, the organism's unique structural unity is premised on the assumption that all the essential parts already exist and are connected together so as to be able to function for the end of self-preservation. Thus, according to Bonnet, the particular *kind* of purposeful functioning found in organic bodies requires that organisms *not* develop epigenetically, part after part.

Albrecht von Haller is another proponent of preexistence who seems especially motivated by phenomena uniquely associated with living beings. He started his career as an anatomist, spending a lot of time examining the structure of organic bodies while also noticing the physiology of living beings, that is, the functioning permitted by the organic structure. When he later turned to the problem of generation, Haller concluded that one could not account for the functioning existence

[18] Charles Bonnet, *Contemplation de la nature,* selections translated by T. S. Hall in *Source Book in Animal Biology* (New York: Hafner Publishing Company, [1769] 1964), 377.

of the parts outside of the whole, and in these later works, he abandoned his early embrace of epigenesis in favor of preexistence.[19] It is crucial to note that, for Haller, the essential fact needing explanation is the nature of living organic bodies as functioning wholes; this biological fact is the starting point for this sort of argument, and it explains why I call this a biological-scientific argument in favor of preexistence, even though the theory of preexistence itself turns decidedly away from natural explanations toward a clearly supernatural one.

Again, it is a further step to say that God created all organisms, fully formed, at the Creation, rather than throughout the duration of the natural world. There are different ways to take this step, but one suggests that God's creating all organisms at the Creation rather than piecemeal upon every new generation avoids the deus ex machina that many early moderns would have wanted to avoid. The crucial point for my purposes is that this way of connecting preexistence with teleology starts with the acknowledgment that there is something unique about living beings, and then proceeds by showing that the only way of accounting for this unique character is through preexistence theory. This can be contrasted with the more straightforward theological use of teleology to argue for preexistence noted previously. That way was through a teleological argument that happens to focus specifically on the organism as the well-designed phenomenon to be explained by appeal to God, but which might have focused (as teleological arguments often do) on any number of other beautifully designed aspects of the natural world. The current argument starts with biological facts about organisms, and then goes on to reach the conclusion the preexistence must hold as the only way of accounting for these uniquely biological facts.

2.3. Metaphysical Argument

The final type of argument in favor of preexistence that I canvas is what I call the metaphysical argument. As noted before, I take this argument to be distinct from theological arguments, even though theological arguments are also metaphysical given their inclusion of God. But metaphysics includes much more besides a consideration of God's existence and nature,[20] and the specific issue I focus

[19] Karen Detlefsen, "Explanation and Demonstration in the Wolff-Haller Debate," in *The Problem of Animal Generation in Early Modern Philosophy*, ed. Justin E. H. Smith (Cambridge, UK: Cambridge University Press, 2006), 235–261.

[20] On the changing nature of metaphysics and the scope of topics that fit within metaphysics, see Alan Gabbey, "Disciplinary Transformations in the Age of Newton: The Case of Metaphysics," in *Between Leibniz, Newton, and Kant*, ed. Wolfgang Lefevre (Dordrecht, NL: Kluwer Academic Publishers, 2001), 3–23; and Gary Hatfield, "Metaphysics and the New Science," in *Reappraisals of the Scientific Revolution*, eds. David C. Lindberg and Robert S. Westman (Cambridge, UK: Cambridge University Press, 1990), 93–166.

on here is the metaphysical issue of individuation. Once again, there is overlap between this and the other categories of arguments. For example, the biological motivation noted before may well be considered an explanation to account for individuation; preexistence is called upon there to explain the unity and singularity of organic bodies. My concern here, however, is with a much more fundamental, specifically a metaphysical, conception of individuation. For at least one thinker—Leibniz—I contend that preexistence is called upon to account for the metaphysics of the individuation of corporeal substances.

Before turning to Leibniz himself, I will make just a few general comments about the metaphysics of individuation as I intend to discuss it here. Raymond Williams, in his *Keywords: A Vocabulary of Culture and Society*, traces the meanings attached to the word "individual." Crucial for my purposes is a contrast that Williams makes between the individual as distinct or different from others of its kind, and the individual as indivisible—that is "... 'an individual'—a single example of a group ... [and] 'the individual': a fundamental order of being."[21] Jorge Gracia captures this contrast in a different way by distinguishing between the concepts of singularity and particularity, and he argues that it is a mistake to equate the individual with both singularity and particularity as if they are one and the same thing. "For particularity has to do with an individual's 'participation in' or 'partaking of' a universal. In this sense the individual is considered as being a part of something else, or as partaking of it.... An individual is called singular ... or is said to have singularity because it is not many."[22] Thus, singularity captures the idea of a being's oneness or unity (in spite of its complexity and through time) in a way that particularity does not.[23] It is this meaning, and therefore the problem of the individual as a fundamental order of being, and not as an example of a group—the problem of the individual as singular, and not the individual as particular—that is central to a consideration of Leibniz's preexistence theory. That is, Leibniz takes the living being to be a paradigmatic

[21] Raymond Williams, *Keywords: A Vocabulary of Culture and Society* (London: Flamingo Fontana Paperbacks, 1983), 163.

[22] Jorge Gracia, *Introduction to the Problem of Individuation in the Early Middle Ages* (München, DE: Philosophia Verlag, 1984), 25–26.

[23] My separation of the broader problem into two sub-problems is not meant to imply that these sub-problems are unrelated, nor is it meant to imply that this is the only way of dividing the broad problem of individuation into distinct concerns. Jorge Gracia provides a sustained conceptual and historical analysis of the various complexities at play in dealing with metaphysical and epistemological problems of the individual. Still, at the outset of his study, Gracia himself divides the problem into these two basic sub-problems (Jorge J. E. Gracia, *Individuality: An Essay on the Foundation of Metaphysics* (Albany, NY: SUNY Press, 1988), 2–3. This twofold division of the problem is found in more recent literature on individuation as well (e.g., E. J. Lowe, "Individuation," in *The Oxford Handbook of Metaphysics*, eds. Michael J. Loux and Dean Zimmerman (New York: Oxford University Press, 2003), 77.

metaphysical individual in the sense of being a paradigmatic example of a fundamental order of being.[24]

A number of commentators believe that Leibniz is motivated to hold the preexistence doctrine for largely the same reasons as his contemporaries, including the biological argument from the unity of organisms noted before.[25] But the usual explanations for embracing preexistence seem less compelling in Leibniz's case because he holds a particular version of that theory that is quite unlike any other version suggested in the 17th century. Certain aspects of his theory indicate that his commitment to mechanism is beside the point, and other aspects of his theory actually *undermine* the functional unity of organic wholes *as we commonly understand them*. Paramount among the unusual aspects of Leibniz's theory is that he maintains that preexistence is equally a theory about 'death' as it is about generation; just as a living body never comes into being, it also never dies (e.g., C 523/AG 34; see also G VI, 620/M §76).[26] No other preexistence theorist holds this, because they all believe that the laws of motion are quite capable of destroying a living body, which is how we experience the death of such a body. I contend that Leibniz's theory of preexistence is not only—maybe not even first and foremost—a theory about generation of new living beings;

[24] Taking the organism as a paradigmatic case of a singular individual has a long history. Aristotle, e.g., has a number of conceptions of what counts as a primary individual substance—what is, in some sense, indivisible. According to one conception, found in *Metaphysics*, book vii, individual substances are unities in the sense that they are intrinsic unities—they are a complex of features essential to each other, none of which can be considered a mere predicate of the others, and none of which can therefore exist apart from the rest (*Meta* vii, 4; 1030a3–6). If anything can be counted among true substances, it is the living being, "a man or a plant or one of the things of this kind" (*Meta* vii, 7; 1032a19–20), a being marked in part by its ability to produce another of its kind (*Meta* vii, 8; 1033a29ff.). Citations from Aristotle, *The Complete Works of Aristotle*, 2 vols., various translators, ed. Jonathan Barnes (Princeton, NJ: Princeton University Press, 1995). *Metaphysics* abbreviated as *Meta*; *Parts of Animals* cited as *PA* and identified by Bekker's 1831 texts by page and line number.

[25] See, e.g., Donald Rutherford, *Leibniz and the Rational Order of Nature* (Cambridge, UK: Cambridge University Press, 1995), 201–203.

[26] All references to Leibniz's works are to the following editions with their respective abbreviations: G. W. Leibniz, *Philosophical Essays*, eds. and trans. Roger Ariew and Daniel Garber (Indianapolis, IN: Hackett Publishing Company, 1989), cited as AG with page number; Leibniz, *Opuscles et fragments inédits de Leibniz*, ed. Louis Couturat (Paris: reprint; Hildesheim, DE: Georg Olms, [1903] 1966), cited as C with page number; Leibniz, *Correspondence with Arnauld*, trans. George Montgomery (La Salle, IL: Open Court Publishing Company, 1995), cited as CA with page number; Leibniz, *Die philosophischen schriften von Gottfried Wilhelm Leibniz*, 7 vols., ed. C. I. Gerhardt (Berlin: Weidman; reprint, Hildesheim, DE: Georg Olms, [1875–1890] 1965), cited as G with volume and page number; Leibniz, *Philosophical Papers and Letters*, trans. and ed. Leroy E. Loemker (Dordrecht, NL: D. Reidel Publishing Company, 1956), cited as L with page number; Leibniz, *Monadology*, trans. George Montgomery (La Salle, IL: Open Court Publishing Company, 1995), cited as M with paragraph number; and Leibniz, *Leibniz's New System and Associated Contemporary Texts*, eds. R. S. Woolhouse and Richard Francks (Oxford, UK: Clarendon Press, 1997), cited as NS with page number.

it is also—maybe first and foremost—a theory about the metaphysics of individuation. Focusing on the metaphysics of individuation helps make sense of the peculiar features of his unusual version of the theory. The bare sketch of the argument that would need to be made to maintain this claim is as follows.

First, Leibniz fairly early identifies the need for a strictly indivisible, indestructible, active, and infinitely complex substantial form in order to account for the unity, and therefore reality, of any substance (e.g., CA 135).[27] His concern here is with metaphysical individuation. Second, he claims that a substantial form is never found without a body (e.g., CA 230; cf. G VI, 619/M §§70 and 72; "New System": G IV, 480–481/NS 14; notes for a letter to Des Bosses, February 5, 1712: G II, 438ff./AG 199ff). Here, I don't think the argument needs to settle on whether that body is metaphysically real, or merely phenomenal, just that for every substantial form, there is a material counterpart always associated with it. Third, Leibniz settles on a living animal body that is infinitely complex as the appropriate material counterpart for the substantial form as he conceives it (e.g., G II, 118/CA 221; "First Explanation of the New System": NS 48; letter to de Volder, June 20, 1703: G II, 251/AG 176). Why? I think the reason is that, like the substantial form, the organic body as Leibniz conceives it is infinitely complex. It is also a true unity in the sense that Justin Smith (and Robert Sleigh, upon whom Smith draws) both indicate (i.e., it is not deconstructible component-wise).[28] Even though it is divisible, it is indestructible (like the substantial form) because no matter how much it is divided, an infinitely complex machine still survives. Like the substantial form, it is internally active (e.g., G II, 77/CA 162; letters to de Volder, especially November 10, 1703: G II, 257–259; and January 21, 1704: G II, 262–265). Fourth, because this animal machine is always associated with a substantial form, and because substantial forms were created by God at the beginning of the universe and will endure until the annihilation of the universe, the bodies associated with them must have been created by God at the Creation and they will also endure until the annihilation of the universe; they neither generate naturally nor die naturally.

And thus we reach Leibniz's own peculiar theory of preexistence, but it is motivated by a need to account for metaphysical individuals. This account also explains the feature of his theory I noted before, namely, preexistence for Leibniz is equally a theory about apparent death as it is about generation. It has to be because it is primarily a theory meant to account for the endurance of individual substantial forms and their bodies.

[27] I focus primarily on Leibniz's mid-1680s correspondence with Arnauld for evidence for my claims here, since this correspondence is where Leibniz's position as I see it emerges with full force.

[28] Justin E. H. Smith, *Divine Machines: Leibniz and the Sciences of Life* (Princeton, NJ: Princeton University Press, 2011), 110–115; R. C. Sleigh, *Leibniz and Arnauld: A Commentary on the Correspondence* (New Haven, CT: Yale University Press, 1990), 126.

If this account is right, then Leibniz's theory of preexistence is not motivated entirely (or even primarily) by a need to account for the generation of living bodies, even though it solves *that* problem too. Leibniz himself makes this order of thinking clear on at least two occasions. In a letter to Foucher regarding the doctrines of the "New System," Leibniz indicates that preexistence follows from the incorruptibility, and thus unity, of the soul (G I, 391/NS 54–55). Similarly, in the "Considerations on Vital Principles," after having explained that substances are autonomous from each other and from God, and are imperishable because of the need to keep them autonomous (G VI, 540ff./L 587–588), Leibniz notes that this issue of individual independence and unity distinct from others and from God has led him to his theory of generation by preexistence (G VI, 543–544/L 589). If the theory were meant merely to account for generation, it would surely be much simpler and more closely aligned with the versions offered by his contemporaries; it would not, for example, be equally a theory of 'death' as a theory of 'generation.'

3. Malebranche on Generation: Mechanism, Biology, and Theology

Malebranche's motivation to endorse preexistence includes all of the reasons just noted except one. I do not believe that he has the sorts of concerns with individuation in the rarified metaphysical sense (as opposed to the biological sense) that we find in Leibniz; indeed, I believe Leibniz alone among the early moderns links the metaphysics of individuation with the theory of preexistence. What is especially interesting about Malebranche is that in at least one place in his corpus, three arguments come together to jointly offer support for preexistence, even though none separately provides an especially powerful argument in favor of this theory of generation. That is, we find in Malebranche a sort of "all things considered" best theory, a theory in which a cluster of commitments typical of a 17th-century thinker can all cohere to give rise to the theory of preexistence. I turn to these three intertwined arguments after briefly noting the role original sin might play in Malebranche's theory of preexistence. I conclude with some thoughts on the status of the life sciences in Malebranche's work.

Malebranche has an easy route to using original sin as evidence in favor of the ovist theory of preexistence. He argues that one way in which original sin can be transmitted to generations beyond the first is through bodily communication, specifically during generation of new individuals (ST 120; ST 195). Specifically because women's bodies contribute so much to the generation of offspring, sin is transmitted by the woman in generation (ST 600). From here, it is an easy step to the original sin motivation in favor of preexistence: Since we are all stained by

Eve's original sin (per the Bible), one might reasonably conclude that the reason for this is that we were all present in Eve's ovaries at the moment of that sin.

I now turn to the passage in the *Search After Truth* in which the teleological-theological, mechanistic-scientific, and biological-scientific arguments come together to provide united support for the likelihood, in Malebranche's estimation, that preexistence is the correct theory of generation:

> ...this thought [that God had formed the world all at once...in the same order, and with the same arrangement of parts that it would have had if He had formed it gradually in the simplest of ways] is worthy of the power and wisdom of God: of His power, since in an instance He made all His works in their greatest perfection; of His wisdom, because...the order of nature could not subsist if the world had been produced in a manner contrary to the laws of motion by which it is preserved....
>
> Furthermore, there is a great difference between the formation of living and organized bodies, and that of vortexes of which the universe is composed. An organized body contains an infinity of parts that mutually depend upon one another in relation to particular ends, all of which must be actually formed in order to work as a whole. For it need not be imagined with Aristotle that the heart is the first part to live and the last to die. The heart cannot beat without the influence of the animal spirits, nor can these be spread throughout the heart without the nerves, and the nerves originate in the brain, from which they receive the spirits. Moreover, the heart cannot beat and pump the blood through the arteries unless they, as well as the veins that return the blood to it, are already complete. In short, it is clear that a machine can only work when it is finished, and that hence the heart cannot live alone.... It would be wrong then to pretend to explain the formation of animals and plants and their parts, one after the other, on the basis of the simple and general laws governing the communication of motion; for they are differently connected to one another by virtue of different ends and different uses in the different species. But such is not the case with the formation of vortexes; they are naturally born from general laws.... (OC II, 343–344/ST 465–466)

I will show how this passage presents an all-things-considered, coherence-of-multiple-rationales argument in favor of the theory of preexistence by discussing each of the three argumentative strands in turn.

First, Malebranche employs the teleological-theological argument for preexistence. This is clear in the first paragraph of the passage cited: That God created the world "with as much perfection as it now has" (ST 464) is evidence for the wisdom and power of God. That is, these features of the world provide evidence for the existence of a being with these features—the core of the teleological argument for God's existence. This argument occurs in the midst of Malebranche's defense of preexistence, the bulk of which occurs after this passage (as we see in the extended quotation above), but which begins before this teleological argument is offered. So, preexistence here is motivated by the teleological argument for God's existence—a theological form of argument.

Three important points should be noted. First, this is clearly an example of the theological argument (and not the biological-scientific argument that I shortly consider), since even though Malebranche offers this argument in the middle of a discussion of the generation of living bodies, the argument itself clearly welcomes the conclusion that *all perfection on the earth*, and not just the perfection of living beings, supports the conclusion that God as a powerful and wise being is the cause of those effects. This argument is motivated by the desire to prove the existence of God, and it relies upon the particularly noticeable perfection of living beings to provide this proof of God's existence. However, this argument does not draw a strong distinction between the living and the nonliving as perfect effects requiring reference to God as cause. Even nonliving things can be used in a teleological argument for God's existence. Nonetheless, the conclusion includes the claim that God produced *all perfection* at the Creation, and this leads to the further conclusion that all living things must have been fully formed, and thus perfect, when God created the world. Second, Malebranche suggests a separation between the metaphysical truth of the natural world (that God created it with all its perfection at the Creation, including the fact that organisms are all fully formed by God at that moment), and how we might study the world in order (as if we were naturalists) to better understand its (God-given) nature. That is, we may well study the generation of living things as "gradually formed in the wombs of their mothers" (ST 465) to better understand how God himself used his principles to create them in the first place. The metaphysical truth, as Malebranche sees it, that organisms are all preformed by God at the Creation, does not preclude the scientist from investigating organisms empirically to better understand their natures. Third, this second point—that we could presumably study the formation of organisms within the wombs of their mothers—shows why this teleological-theological argument in favor of preexistence cannot stand alone. For if we could give a complete and satisfactory account of the gradual formation of the fetus, while still maintaining the perfection of God as its cause, then the teleological-theological argument cannot preclude preexistence's prime rival, the theory of epigenesis. Only by turning to the two scientific arguments found in this passage can we reject epigenesis.[29]

So Malebranche includes an acknowledgment of the particular perfection of living beings in his teleological-theological argument for God's existence, thus demonstrating that he has this motivation for embracing preexistence, but he is also committed to the two scientific arguments I have identified—arguments that we can attribute to him given his openness previously noted to scientific inquires into metaphysical facts about our world. Malebranche is clearly committed to mechanism, but he also clearly believes that mechanism cannot account for the existence of living beings (the mechanistic-scientific argument). Moreover, he is led to the conclusion that mechanism cannot account for the existence of living beings because of the unique feature of living bodies (the biological-scientific argument). I will treat these in turn.

Malebranche is committed to mechanism, both the form that suggests that the world and its parts are to be thought of as machines, and the form that suggests that all phenomena are the result of sub-visible parts of matter interacting and colliding in accordance with lawful motion (micro-mechanism). The explicit parallel between a living body and a machine in the long passage previously quoted establishes his commitment to machine mechanism. His discussion of the hypothetical formation of plants and animals due to simple and general laws of the communication of motion from body to body establishes his commitment to micro-mechanism. Moreover, it is his refusal to abandon micro-mechanism that leads him to preexistence: For it would be "wrong... to pretend to explain the formation of animals and plants and their parts, one after the other" in terms of micro-mechanism. Thus his refusal to abandon mechanism—his embrace of it—leads him to preexistence.

But the question arises: Why *cannot* micro-mechanism explain the formation of living bodies? And here Malebranche turns to the biological-scientific argument in order to supplement the mechanistic-scientific argument, which alone cannot provide an answer to that question—and so cannot bring us to the conclusion of preexistence. The reason why mechanism cannot explain the formation of living bodies is that a living body—unlike any other body—has characteristics unique to it. It is the recognition of the uniqueness of living bodies that makes this argument a *biological-scientific* argument, in distinction from

[29] A fourth, extremely interesting, historical point, which is not central to my line of argument here, is that evident here is a conclusion about the teleological argument that will eventually be made explicitly by Hume, namely, even if we grant that the teleological argument can establish the existence of God, all it can establish about God is that he is wise and powerful, but not that he is benevolent or good. See David Hume's *Dialogues Concerning Natural Religion*, part XII.

the teleological-theological argument that may recognize the extreme perfection of living bodies, but can also recognize the perfection of all of God's handiwork. The uniqueness, for Malebranche, of living bodies comes clear when he grants that mechanism *can* explain some forms of organization (vortices) but not other forms of organization (living bodies). It is the unique functioning wholeness of living beings that Malebranche underscores when he writes of the "parts that mutually depend upon one another in relation to particular ends, all of which must be actually formed in order to work as a whole." There are two senses of "purpose" at work in Malebranche's discussion of living bodies. First, there are *ends* served by parts suitably structured to *function* toward those ends. And second, the *whole* is sustained and preserved as a whole due to the proper functioning of the parts whose good function, in turn, is permitted by the *preservation of the whole*. The crucial step in his argument from the recognition of the unique, teleological unity of the organism to the conclusion that they must have always existed as such is the step from recognizing the *ontological* priority of the whole to the parts, to asserting a *temporal* priority of the whole to the parts. That is, since parts cannot live outside the whole, Malebranche concludes that the whole must have always been there to allow the parts to function and therefore live.

This final point is worth some expansion. Some preexistence theorists are motivated, even if they do not explicitly recognize this, by concerns about living beings that are essentially Aristotelian. As Aristotle argues, organs and other body parts outside of an organism cannot live or function and are therefore *different in kind* from those living, functioning parts found connected in a living whole. The point is underscored in the *Parts of Animals*:

> And yet a dead body has exactly the same configuration as a living one; but for all that is not a man. So also no hand of bronze or wood or constituted in any but the appropriate way can possibly be a hand in more than name. For like a physician in a painting, or like a flute in a sculpture, it will be unable to perform its function. Precisely in the same way no part of a dead body, such I mean as its eye or its hand, is really an eye or a hand. (*PA* i, 1; 640b35–41a5)

George Cuvier, for example, believes that the successive formation of the organism, part after part, implies the death of the individual *before* life even begins because parts outside of a whole are not living.[30] Other preexistence theorists, as

[30] W. Coleman, in writing on Cuvier's embrace of preexistence, states this idea nicely. For Cuvier, "successive development of the various parts would produce only physiological chaos and the death of the individual"; Coleman, *George Cuvier: Zoologist* (Cambridge, MA: Harvard University Press, 1964), 128. Of course, for Aristotle, the organism forms part after part, but the *whole* nonetheless preexists in the form that is passed from male to female.

we saw in the passage cited from Bonnet, and as we find in the current passage from Malebranche under discussion, are acutely aware of the unity—the oneness and wholeness—of living bodies, and the related fact that organic parts can function only inside a whole and, indeed, lose their characteristic nature as *living* parts outside of the whole. So, many preexistence theorists hold a similar conception of living beings that we find in Aristotle, the one that drives his criticisms of the part-after-part generation theories of his materialist predecessors such as Empedocles and Hippocrates. And many preexistence theorists hold this conception because they are either working scientists, like Haller,[31] or are alert to the findings of working scientists, like Malebranche,[32] and therefore acknowledge (like Aristotle) the empirical facts of fully formed organisms, facts that thus constrain the theory of generation that must account for these facts.

Still, many preexistence theorists are also materialists—and (perhaps) mechanists—with respect to nonhuman creatures, including living beings. So their materialist and mechanist commitments preclude an Aristotelian *ontological explanation* for their basically Aristotelian conception of living beings. Instead, they turn to a roughly Platonic account (i.e., God created organisms as unified wholes). Such organisms *cannot* be constructed as we construct a clock, for example. What we do when building a clock is to put together parts into a working whole in which the parts retain their original character. Living bodies cannot be built up out of preexisting parts in this fashion. That is why God must create them *miraculously* at the Creation. This, at least, is one major motivating factor in Malebranche's account of preexistence

Still, the conclusion that the whole must have always been there to allow the parts to function and therefore live does not follow from the biological-scientific argument. One could easily counter that, indeed, the whole cannot live without all the parts in place, interconnected and functioning, but that nonetheless, epigenesis is the correct theory, and the living being simply does not become alive until all the parts have finally formed and can work together as a whole. Here, the co-presence of the teleological-theological argument together with the scientific arguments could make the case for preexistence. As Malebranche writes in a theological mood: "God created the world with as much perfection as it now has," and fully functioning, living organisms are among the most perfect expressions of God's wisdom and power. So, in choosing between preexistence and epigenesis, the teleological-theological argument can help make the case that preexistence best explains the uniqueness—indeed unique perfection—of living bodies.

[31] Detlefsen, "Explanation and Demonstration."

[32] Detlefsen, "Supernaturalism, Occasionalism, and Preformation."

All three strands of argumentation—teleological-theological, mechanistic-scientific, and biological-scientific—are present in this long passage in Malebranche's *Search After Truth* in which he defends the ovist version of pre-existence. Each of these three strands is too weak to convince the reader of the correctness of this theory of generation. But the three arguments working together do provide a fairly strong case for preexistence should the reader accept Malebranche's theological and mechanistic commitments and be working with the level of biological knowledge he had. That is, his argument for preexistence here is more of an "all things considered" approach to provide strong evidence (but not decisive arguments) for preexistence as the most likely theory of generation. At the same time, we can see in this argument a combination of some central 17th-century concerns and commitments. First is the commitment to mechanism, both machine mechanism and micro-mechanism, in explanation of natural phenomena. Second is the completely compatible commitment to teleology in the natural world. This teleology takes the form of God's purposes leading him to create natural machines such that they will be able to fulfill his purposes, including his purposes with respect to the parts of organisms and organisms as self-preserving wholes. It does not take the more Aristotelian form of a principle of change internal to bodies and striving toward ends intrinsic to those bodies—a form of teleology indeed at odds with mechanism. This teleology as manifest in the machines of the world permits both an important argument for God's existence and allows the recognition of the unique qualities of living bodies, thus indicating an alertness to the special character of living beings, an alertness that is very promising for a science of life even within this age of committed mechanists. So in this argument in Malebranche's *Search*, we see just why a smart mind of the 17th century would embrace preexistence, that the reasons go beyond the limits of mechanism, and that the reasons open up rich territory for scholars keen on understanding the importance of the life sciences in the 17th century.[33]

[33] This point is quickly becoming unnecessary to make. Historians of science displayed a much earlier openness to a study of the life sciences in the 17th century. See, e.g., Adelmann, *Marcello Malpighi*; Bodemer, "Embryological Thought"; B. Richard, *Descartes' Medical Philosophy: The Organic Solution to the Mind-Body Problem* (Baltimore, MD: Johns Hopkins University Press, 1983); Gasking, *Investigation*; T. S. Hall, *Ideas of Life and Matter*, 2 vols. (Chicago: University of Chicago Press, 1969); Oppenheimer, *Essays*; Roe, "The Development" and "Rationalism and Embryology"; and C. U. M. Smith, *The Problem of Life: An Essay in the Origins of Biological Thought* (New York: John Wiley & Sons, 1976). Historians of philosophy on the continent have also been attuned to the life sciences in this period. See, e.g., Georges Canguilhem, *La formation du concept de réflexe aux XVIIe et XVIIIe siècles* (Paris: Presses Universitaires de France, 1955); Canguilhem, *La Connaissance de la Vie*, 2nd ed. (Paris: J. Vrin, 1969); Canguilhem, *Idéologie et rationalité dans l'histoire des sciences de la vie: nouvelles études d'histoire et de philosophie des sciences* (Paris: J. Vrin, 1977); Annie Bitbol-Hespériès, *Le principe de vie chez Descartes* (Paris: J. Vrin, 1990); Marie-Noëlle Dumas, *La pensée de la vie chez Leibniz* (Paris: J. Vrin, 1976); Geneviève Rodis-Lewis, "Limitations of

Acknowledgments

Many thanks for extremely fruitful discussions with audiences at The Life Sciences in Early Modern Philosophy: A Workshop (Princeton, NJ, May 2011); New York/New Jersey Workshop in Early Modern Philosophy (John Jay College, NY, May 2010); and the American Philosophical Society Eastern Division Annual Meeting (New York, December 2005). This material is based on work supported by the National Science Foundation under Grant no. 0432156.

the Mechanical Model in the Cartesian Conception of the Organism," in *Descartes: Critical and Interpretive Essays*, ed. Michael Hooker (Baltimore, MD: Johns Hopkins University Press, 1978); and Vincent Aucante, *La Philosophie médicale de Descartes* (Paris: Presses Universitaires de France, 2006). Only more recently, however, has interest in early modern life sciences in the English-speaking world surged. See, e.g., Dennis Des Chene, *Spirits and Clocks: Machine and Organism in Descartes* (Ithaca, NY: Cornell University Press, 2001); Daniel Fouke, "Mechanical and 'Organical'"; Fouke, "Spontaneity and the Generation of Rational Beings in Leibniz's Theory of Biological Reproduction," *Journal of the History of Philosophy* 29.1 (1991): 33–45; Stephen Gaukroger, "The Resources of a Mechanist Physiology and the Problem of Goal-Directed Processes," in *Descartes' Natural Philosophy*, eds. Stephen Gaukroger, John Schuster, and John Sutton (New York: Routledge, 2000), 383–400; Gaukroger, *Descartes' System of Natural Philosophy* (Cambridge, UK: Cambridge University Press, 2002), chapter 7; Gary Hatfield, "Descartes' Physiology and its Relation to his Psychology," in *The Cambridge Companion to Descartes*, ed. John Cottingham (Cambridge, UK: Cambridge University Press, 1992); Justin Erik Smith, "On the Fate of Composite Substances after 1704," *Studia Leibnitiana* 30.2 (1998): 204–210; Justin E. H. Smith (ed.), *The Problem of Animal Generation in Early Modern Philosophy* (Cambridge, UK: Cambridge University Press, 2006); Smith, *Divine Machines*; Catherine Wilson, *The Invisible World: Early Modern Philosophy and the Invention of the Microscope* (Princeton, NJ: Princeton University Press, 1995); and Wilson, "Leibniz and the Animalcula," in *Studies in Seventeenth-Century European Philosophy* (Oxford, UK: Clarendon Press, 1997). See also, as just a few notable examples of his extensive work on the philosophy of early modern life sciences, the following by François Duchesneau: "Du Modèle Cartésien au Modèle Spinoziste d l'etre Vivant," *Canadian Journal of Philosophy* 3.4 (1974): 539–562; "Malpighi, Descartes, and the Epistemological Problems of Iatromechanism," in *Reason, Experiment and Mysticism in the Scientific Revolution*, eds. M. L. Righini Bonelli and William R. Shea (New York: Science History Publications, 1975); "Leibniz et la théorie physiologique," *Journal of the History of Philosophy* 15 (1976): 281–300; "Haller et les theories"; *La Physiologie de Lumieres Empirisme, Modeles et Theories* (The Hague, NL: Martinas Nijhoff Publishers, 1982).

9

Réaumur's Crayfish Experiments in Hartsoeker's *Système*

Regeneration and the Limits of Mechanism

CATHERINE ABOU-NEMEH

1. Growing Doubt: 1712 Crayfish Regeneration

In 1712, René-Antoine Ferchault de Réaumur[1] (1683–1757) shocked the scientific community with the findings of his crayfish experiments. Réaumur's extensive observations confirmed spontaneous regeneration of amputated crayfish claws. When Réaumur severed the leg of a crayfish at the joint, the leg grew back.[2] Although Réaumur was not the first to report about the regenerative capacities of crayfish, his lengthy observations built on those of Jean-Baptiste Du Tertre (1610–87) and circulated more widely.[3] Du Tertre had written about regeneration in crabs in his travelogue on the Antilles that was published first in 1654 but did not receive much attention until Réaumur revived the topic.

[1] Réaumur came to Paris in 1703 for the first time, and in 1708 was nominated by Pierre Varignon to be his "student geometer" at the Académie des Sciences. See J. B. Gough, "Réaumur, René-Antoine Ferchault De," in *Complete Dictionary of Scientific Biography*, vol. 11 (Detroit, MI: Charles Scribner's Sons, 2008), 327–335, 328. *Gale Virtual Reference Library*, Web. 2, Dec. 2011.

[2] René Antoine Ferchault de Réaumur, "Sur les diverses Reproductions qui se font dans les Écrevisses, les Omars, les Crabes, etc. et entr'autres sur celles de leurs Jambes & de leurs Écailles," in *Histoire de l'Académie Royale des Sciences, Avec les Mémoires de Mathématique & de Physique pour la même Année, Tirés des Registres de cette Académie*, année 1712 (Paris: L'Imprimerie Royale, 1731), 223–241; hereafter *HARS*. See also "Sur la Reproduction de quelques parties d'écrevisses," *HARS*, 34–35 ; and furthermore, Jacques Roger, *The Life Sciences in Eighteenth-century French Thought*, trans. Robert Ellrich (Stanford, CA: Stanford University Press, 1997), 347.

[3] Jean Baptiste Du Tertre, Histoire generale des isles des Christophe, de le Gvadelovpe, de la Martiniqve et avtres dans l'Ameriqve, où l'on verra l'establissement des Colonies Françoises, dans ces Isles; leurs guerres Ciuiles & Estrangeres, & tout ce qui se passe dans les voyages & retours des Indes (Paris: Jacqves et Emmanvel Langlois, 1654), 377–378.

Its revival had unintended consequences: News of Réaumur's crayfish experiments in Paris transformed Nicolas Hartsoeker's system-building. He professed to have converted his philosophical beliefs after his patron, Johann Wilhelm II of the Palatinate in Düsseldorf, told him of Réaumur's crustacean trials in 1712. Hartsoeker did not take Réaumur's experiments at face value; he repeated the experiment on more than 200 crayfish, always getting the same results as Réaumur. Time after time, the crayfish would regrow their missing legs. In a 1722 article in the *Bibliothèque ancienne et moderne*, Hartsoeker corroborated Réaumur's findings: "I deemed this experiment conclusive against those who claim that in the beginning God has created all plants, trees and animals that have already been and that will be in the centuries to come, as if he had encased them one in the other."[4] On the basis of regeneration, our Dutchman rejected the theory of encasement/preformation (*emboîtement*) and thus made room for his own formative principle. Before he laid out his new explanatory scheme, he considered two alternative scenarios to account for the crayfish trials that he quickly rejected: (1) that we live in a universe where God makes everything continually, "as if by his own hands," and (2) that a universe is a great machine where everything moves according to a prime mover that God triggered in the very beginning, depending on "a mere mechanism." In both these plausible but ultimately flawed alternatives, mankind would be deprived of freedom and free will. Hartsoeker opted for a third explanation: An intelligent force, or soul, was responsible for the repair of the leg. He explained his concept of "Intelligences" as "subordinate to God" but constantly at work in the production and conservation of living things. These Intelligences were crucial intermediaries between inert matter particles and God, and were part of a hierarchy of more and less intelligent beings that enlivened organic and inorganic matter.[5]

In what follows, I explore the role of the 1712 crayfish experiments in Hartsoeker's construction of a workable system of natural philosophy. Significantly, Hartsoeker had begun his philosophical turn at least 16 years before hearing of Réaumur's crayfish findings. That is, since 1696, he had gradually begun reforming his Cartesian-inspired cosmology to include Neoplatonist ideas on the soul. I argue that the crayfish trials allowed Hartsoeker to further develop and assert his proposition that Intelligences mediated between inert matter and God. In so doing, I discuss the relationship between Hartsoeker's Remonstrant milieu and several English philosophers, and suggest the ways in

[4] Nicolas Hartsoeker, "Lettre de Mr. Hartsoeker à l'Auteur de la Bibliothèque A. & M. sur les Serres, qui recroissent aux Ecrevisses, quand on les a rompues, sur la petitesse des Animaux qui quelques-uns supposent avoir été tous créez au commencement du Monde & sur les Natures, qui forment présentement les Corps Organizez, & qui y résident," *Bibliothèque ancienne et moderne* 18 (1722): 195.

[5] Ibid., 200–201.

which his immediate family and intellectual circles influenced his ideas about free will and the soul. Crucial to this development were Hartsoeker's intellectual debts to the Cambridge Neoplatonists Henry More and Ralph Cudworth. Their arguments for a metaphysics of physics persuaded Hartsoeker to begin modifying parts of his system. In a published article, Hartsoeker claimed that More's arguments in his letters to Descartes convinced him to begin looking outside the Cartesian framework for answers. Meanwhile, Cudworth's idea of "Plastick Nature," at least in part, inspired Hartsoeker's "organizational" first element that imbued matter particles with motion in his revised system of physics. This chapter discusses the philosophical traditions within which Hartsoeker was developing his cosmology. It also illustrates the reception of Cambridge Platonism in continental Europe and the challenges to free will that cosmologies like Descartes's continued to pose.

Hartsoeker applied his interpretation of the crayfish trials to general natural phenomena, such as regenerative faculties and survival instincts in humans and animals. These hitherto remained without a causal explanation. For instance, it was the Intelligence residing in Hartsoeker's body that was responsible for healing him when he had cut himself: "It is only the Intelligence that lives in my body that cures me by repairing the cut veins. . . ."[6] What, if not an Intelligence at work, gives an infant the instinct to suckle at its mother's breast? What, if not an Intelligence, induces a baby spider to spin a web to catch flies?[7] He not only explained the basic instincts of preservation in this manner, but motion too. The intelligent force that resided in a body governed it and "impart[ed] it with or deprive[d] it of movement."[8]

Before the crayfish experiment made him think otherwise, Hartsoeker had believed that preexistent germs—absorbed into the body with air or food—nourished, replenished, and developed the body from within. Réaumur's crayfish experiment saw him rejecting "so absurd and bizarre a thought."[9] Hartsoeker discarded both Cartesian mechanism and the preexistence of germs as viable explanatory theories. Instead, he considered the possibility that "the mechanical formation of a single claw, or leg, is not less inconceivable, not less impossible," than that of a complete animal. Hence, "it is no less absurd to think that God would make a new creation out of it. Thus we must, by absolute necessity, have

[6] Ibid., 202.

[7] Ibid., 203.

[8] Ibid., 204–205.

[9] Nicolas Hartsoeker, "Remarques sur une These de Physique que Mr. Muller, Professeur en Philosophie etc. à Léipsic, a fait soutenir par Mr. Platner sur la Generation des Animaux, à l'occasion du septième Discours de la Suite de mes Conjectures Physiques, où je parle de cette generation," *Recueil de plusieurs pieces de physique* (Utrecht, NL: la Vueve de G. Broedelet & Fils Libraires, 1722), 193.

recourse... to an intelligence... that dwells in this animal and repairs the loss, as soon as it happens."[10] This Intelligence seemed to govern the entire animal and control its behavior.

What corroborated Hartsoeker's invention of an Intelligence was the crayfish's own seemingly conscious role in the amputation. Réaumur explained that the fullest regeneration happened when the pincer was severed right above the fourth articulation, where the forelimb is slenderest. And since crayfish bled very freely, casting off the limb or amputating it voluntarily at the narrowest point, or upper joint, would be the safest way to stave off a fatal hemorrhage. At times, crayfish would lose one of their limbs during their seasonal moult, or shedding, of their carapace. The shedding process stretched over (usually three) days and could be fatal to the crayfish. Not infrequently, the exhausting operation of the moult resulted in a limb being torn off along with the shell.[11] Geoffroy the Younger, for instance, believed the crayfish was weak and sickly during this time, and posited that a "disease... causes the moult" in crayfish.[12] If injured at any lower joint, the crayfish allegedly maimed itself above the fourth articulation on purpose to escape its captor alive and have its limb regrow in full.[13] Such a response in the crayfish was exemplary of natural—if not, divine—providence. It was as if the crayfish instinctively knew how nature could help it remedy a potentially horrible injury.

But why then didn't humans regrow their limbs after an amputation? Hartsoeker categorized human amputations as relatively rare, chance occurrences. Therefore, like Réaumur, he reasoned by invoking the principle of necessity: "... and perhaps if some of our limbs were also as easily torn off as those of crayfish, we would have received from Nature the same advantage as these animals."[14] Crucially, although Hartsoeker fell back on the theory of informed matter to explain regeneration, Réaumur offered the theory of preexistence

[10] Nicolas Hartsoeker, "Lettre sur quelques endroits des ouvrages des Mrs Cheyne et Derham, à l'auteur de la Bibliothèque ancienne et moderne," *Bibliothèque ancienne et moderne* 8 (1717): 336. In a letter to Jean Le Clerc that was published in the *Recueil de plusieurs pieces de physique* (Utrecht, NL, 1722), Hartsoeker once again revisited the topic of regeneration in crayfish. In one breath, he rejected the theory that held that animals were encased within animals from the beginning of time and that they were mechanically formed on pages 7–8.

[11] Thomas Henry Huxley, *An introduction to the study of zoology, illustrated by the crayfish* (New York: D. Appleton and Company, 1901), 37–39.

[12] Geoffroy le Jeune, "Observations sur les ecrevisses de riviere," in *Mémoires de l'Académie Royale des Sciences* in *HARS*, 1709 vol. (Paris: L'Imprimerie Royale, 1711), 313.

[13] Réaumur, "Sur les diverses Reproductions qui se font dans les Écrevisses," 228. Nineteenth-century naturalists believed that decapods, such as the crab and lobster, "practice defensive mutilation or autonomy." See Francis Hobart Herrick, *Natural History of the American Lobster* (Washington, DC: Government Printing Office, 1911), 100.

[14] Hartsoeker, "Lettre de Mr. Hartsoeker à l'Auteur de la Bibliothèque A. & M. sur les Serres, qui recroissent aux Ecrevisses...," 202, 226.

of germs as a possible account. The Frenchman speculated that the new limb regrew from little eggs in the crayfish that contained the limb in miniature. The nourishing juices that flowed through a living crayfish then would have triggered that germ to grow and become the new leg (as if from an activated stem cell).[15] In short, Réaumur continued to explain the crayfish experiment with the theory of preexistent germs.

Regeneration in crayfish stupefied another contemporary, Fontenelle, the Parisian Academy's secretary who recorded and commented on Réaumur's findings. He remained more skeptical than Réaumur and did not volunteer a separate theory to explain the regrowth of crayfish limbs. Fontenelle admitted that Réaumur's crayfish experiments undermined the theory of preformation. He carefully reckoned that "[a]lthough the theory that maintains that the animal is already fully formed in the egg explains generation rather well, the phenomenon of generation is still rather marvelous. But that another organic part of an animal similar to the one which has been cut off is reborn, that is a second marvel of a different nature from the first one and cannot be explained by the theory of ovism [*système des Oeufs*]" (preformation).[16] In other words, the theory of preformation explained generation successfully, but seemed to fizzle out when applied to regeneration. Interestingly, Hartsoeker ventured further than Réaumur and Fontenelle in interpreting the crayfish experiment. Fontenelle believed the regrowths in crayfish exemplified "rare phenomena."[17] Hartsoeker, meanwhile, took the extraordinary phenomenon of regeneration in crayfish as illustrative of and proof for intelligent matter. In fact, the crayfish experiments had a profound impact on Hartsoeker's notions of matter and soul, and served to support one of the main principles of his system.

2. Generation Debates and Regeneration

Generation stood at the heart of questions about life. Ideas about why living things were formed and how they grew were hotly contested in the middle of the 17th century. The debate centered on the question of the ensoulment of the seed, and thus the ultimate origin of a living being. In the early 17th century, physicians and thinkers offered an explanation for this conundrum in the theory of preformation of the germ or seed.[18] According to this doctrine, the organism

[15] Réaumur, "Sur les Diverses Reproductions qui se font dans les Écrevisses . . . ," 231–232.

[16] Bernard Le Bovier de Fontenelle, "Sur la reproduction de quelques parties des ecrevisses," in *HARS*, 1712 vol. (Paris: L'Imprimerie Royale, 1714), 35.

[17] Ibid., 37.

[18] The ancient Greek notion of σπέρμα (sperm) carried the larger connotation of seed or germ, applicable to the animal, as well as plant kingdoms.

already existed entirely formed in miniature in the semen of the animal. It developed into a full-fledged living being thanks to the soul it received in the seed of the male parent. This belief supported Aristotle's dualist claim that the male parent conferred a form, or soul, while the female contributed matter, or her body, in producing offspring. Partially overlapping with preformation of the germ, the theory of the preexistence of germs began to circulate around 1670. It also presupposed the enlargement of a preformed organism, but it rejected the male parent as the author of the seed. Instead, the doctrine held that God had created the germ in the beginning of time and the male parent served merely as its host. The paganism inherent in Aristotle's scheme found refuge in Platonic thought, which offered a complex, more nuanced, and perhaps mystical exposition of forms. In addition, the materialism inherent in the mechanical philosophy made room for Neoplatonic, chymical, and spiritualist philosophers to explore how matter organized itself. For instance, Jean-Baptiste Morin believed in a divine immaterial spirit, while Jean-Baptiste van Helmont postulated a seminal principle responsible for the generation of beings.[19]

In the 17th century, spontaneous generation from dead and living matter was corroborated by daily experience and by eminent natural philosophers, such as Mersenne, Harvey, and Descartes. Spontaneous generation, especially of insects, worms, and slugs, was accepted almost unanimously until the last third of the 17th century.[20] It was believed to occur through the providence of God or by chance. Nicolas Hartsoeker, for one, explained monstrous births as nature erring at random and healthy births, conversely, as ordained by God. According to Hartsoeker, two animalcules of semen produced a monster "by chance," if they both entered the same egg.[21] Spontaneous generation—or generation by chance—challenged the notion that beings, such as worms that were born out of putrid matter, were also the work of God. Augustine's "seminal principle" suggested that in order to produce living things, matter must have seed.[22] Despite the possible theological and theoretical arguments, the exact process of spontaneous generation remained unclear. One thing seemed obvious: Not just any matter could generate any kind of being. As in explanations of generation, those of spontaneous generation necessarily invoked some kind of nutritive spirit or soul acting on matter that was necessary for life.[23]

Spontaneous generation was a widely held theory in the first half of the 17th century. One of the reasons for its popularity was its compatibility with Descartes's

[19] Roger, *The Life Sciences in Eighteenth-century French Thought*, 259–261.

[20] Jean Rostand, *La Genèse de la Vie: Histoire des Idées sur la Génération Spontanée* (Paris: Libraire Hachette, 1943), 19.

[21] Nicolas Hartsoeker, *Suite des Conjectures Physiques* (Amsterdam: Henri Desbordes, 1708), 130.

[22] Rostand, *La Genèse de la Vie*, 14.

[23] Roger, *The Life Sciences in Eighteenth-century French Thought*, 60–62.

mechanism, which relied on a set of natural laws that could produce life at any time. Spontaneous generation ran out of steam, though, by the end of the century with the minute observations of Hooke, Leeuwenhoek, Malpighi, and Swammerdam. These microscopists brought to light the highly ordered, complex structures of organic life. The newly unveiled microcosms were more neatly explained with theories such as preformation and the preexistence of germs.[24] Regeneration in animal species, on the other hand, was largely ignored by natural philosophers of the day.[25] In antiquity, Aristotle had indeed described how the eyes of very young swallows would regenerate and "sprout afresh" after having been "put out."[26] And Pliny the Elder had told of how lizards and serpents would regrow their lost tails.[27] Their observations were revived and corroborated by humanist natural historians. Overall, however, regeneration was thought to occur almost exclusively in the plant kingdom.[28]

Regeneration of organic parts was well established in trees, whose cut-off branches would grow back and whose ability to adopt and nourish a foreign tree shoot was exploited in the practice of grafting that was already known in ancient times.[29] "But the parts lost by Animals are lost forever, & they would definitely not get a new [body] part that would be grafted [*grefferoit*] onto them," wrote Fontenelle. "Only the legs of Crayfish are reborn, but this privilege is peculiar to them; and should it extend to other Animals, it will always be rather unusual."[30] When Réaumur first sought to explain regeneration of crayfish limbs in 1712, he and his contemporaries like Fontenelle understood the phenomenon in the context of early 18th-century debates on reproduction and rebirth.

The crayfish's annual habit of shedding its carapace was yet another, though far less atypical, example of its unique regenerative capabilities. Rather than couch the phenomenon of limb regrowth within regeneration, Réaumur spoke of "the reproduction of legs and shells" of crayfish.[31] Interestingly, Hartsoeker took Réaumur's

[24] Alain F. Corcos, "Fontenelle and the Problem of Generation in the Eighteenth Century," *Journal of the History of Biology* 4 (1971): 364–365.

[25] Corcos, "Fontenelle and the Problem of Generation," 370.

[26] Aristotle, *De Generatione animalium* in *The Works of Aristotle*, vol. V, trans. Arthur Platt (Oxford, UK: Clarendon Press, 1912), book IV.6, 774b31–34.

[27] Pliny the Elder, *The Natural History of Pliny*, trans. John Bostock and H. T. Riley (London: George Bell & Sons, Pocock, 1975), vol. III, ch. 111, 92–93.

[28] Robert O. Becker and Gary Selden, *The Body Electric: Electromagnetism and the Foundation of Life* (New York: William Morrow and Company, 1985), 32.

[29] Bernard Le Bovier de Fontenelle, "Sur la Reparation de quelques parties du Corps humain mutilées," *HARS*, 1719 vol. (Paris: L'Imprimerie Royale, 1721), 29.

[30] Ibid.

[31] Réaumur's title of his first article on crayfish regeneration: "Sur les diverses Reproductions qui se font dans les Écrevisses, les Omars, les Crabes, etc. et entr'autres sur celles de leurs Jambes & de leurs Écailles," *HARS*, 1712. Réaumur, "Additions aux Observations sur la Muë des Ecrevisses, données dans les Memoires de 1712," in *Mémoires de l'Academie Royale des Sciences*, 1718 vol. (Paris: L'Imprimerie Royale, 1741), 263: "...la reproduction qui se fait des Jambes cassées des

and his own repeated experiments to make a case for *regeneration*. He saw that the leading theory of *reproduction*—germ theory—invoked by both Réaumur and Fontenelle proved useless in explaining limb regrowth in crayfish. Why did Hartsoeker venture to do this and seek a new explanatory framework as a result? I argue that his exposure to Cambridge Platonism in tandem with his growing awareness of the shortcomings of the Cartesian system of nature made him sympathetic to non-mechanistic explanations, which he came to adopt over the years.

To identify how Hartsoeker's experimental investigations applied to his theorizing, it is crucial to understand why and how his thoughts had to travel across the English Channel to adopt ideas of a World Soul and a Plastick Nature. Evidently, Cartesian mechanism could not provide a satisfactory explanation of living things. Matter particles in motion could not suggest causes for phenomena like regeneration or animal souls. In Christian Europe, the role of free will had been compromised by the mechanistic, materialist, and seemingly deterministic implications of Descartes's, Hobbes's, and Spinoza's natural philosophies. Still, this factor alone did not warrant Hartsoeker's choice of an active Intelligence that governed matter in his system. As we have seen with Réaumur himself, a plastic nature theory was *not* the only possible interpretation of regeneration.

Hartsoeker was exposed to Cartesian thought in his student days in Amsterdam and at the University of Leiden. From 1678, he spent time at the Royal Academy of Sciences in Paris and, from 1704, at the court in Düsseldorf. [32] However, he began to rethink some of its principles and assumptions in the 1690s along the lines of the Cambridge Platonists.[33] The most principal and prolific authors of this group were Henry More (1614–87) and Ralph Cudworth (1617–88). The English Platonists were receptive to the new Cartesian philosophy because of the primacy it placed on reason. At the same time, they were hostile to the determinism it implied. They believed that there were two sources of arriving at truth: scriptural revelation and human reason.[34] Not surprisingly, the Cambridge Platonists distrusted a physics

Ecrevisses. . . . J'ai parlé dans le même Memoire d'une autre reproduction qui se fait régulierement chaque anné dans ces animaux, c'est celle de leurs Ecailles." And also on p. 272 of this same article: ". . . la reproduction des Jambes & des Ecailles. . . ."

[32] A crucial part of Hartsoeker's intellectual life was spent at the Academy of Sciences in Paris. Numerous academy members held heterodox views. There were many Jansenists, Cartesians, and Spinozists, but few Neoplatonists. See Alice Stroup, "Censure ou querelles scientifiques: l'affaire Duclos (1675–1685)," *Règlement, usages et science dans la France de l'absolutisme*, eds. Christiane Demeulenaere-Douyère and Éric Brian (Paris: Editeur Technique et Documentation, 2002), 451.

[33] Stroup, "Nicolas Hartsoeker, savant hollandais associé de l'Académie et espion de Louis XIV," *Cahiers d'histoire et de philosophie des sciences, n°47: De la Diffusion des sciences à l'espionnage industriel XVe-XXe siècle*, ed. André Guillerme (Fontenay Saint-Cloud, FR: ENS Editions, 1999), 224.

[34] Dominic Scott, "Reason, recollection and the Cambridge Platonists," in *Platonism and the English Imagination*, eds. Anna Baldwin and Sarah Hutton (Cambridge, UK: Cambridge University Press, 1994), 142–143.

without metaphysics.[35] The Cambridge Platonists and other theologian philosophers appeared in various registers over the course of Hartsoeker's life. They figured prominently in his reading and research, and in the immediate intellectual circle of his father Christiaan Hartsoeker (1626–83), a well-known Remonstrant preacher and close friend of the famous theologian Philip van Limborch.

3. The Dutch-English Connection and Reformed Religion

The unique ties between Dutch Remonstrant and English theologians and political thinkers figured prominently in the background to these religious, philosophical, and political debates. Dutch Remonstrants, also named Arminians after the Remonstrant theologian Jacobus Arminius, faced religious persecution at home. They comprised a small elite of theologians and intellectuals who were politically influential. Sir William Temple, an English ambassador in The Hague, described what distinguished the Remonstrants from their reformed brethren in an account of his travels to the United Provinces in the late 1660s and early 1670s: "The *Arminians*, though they make a great Name among them, by being rather the distinction of a Party in the State, than a Sect in the Church; yet are, in comparison of others, but few in number, though considerable by the persons, who are of the better quality, the more learned and intelligent men, and many of them in the Government."[36] They wrote letters to English theologians and representatives of the Church of England, including Henry More; and they lobbied for support from a much larger (although not less religiously volatile) country on their behalf. And so it was vice versa. On the whole, the works of Colie, Simonutti, Zijlmans, Jardine, and other historians demonstrate the ways in which the Dutch and the English shared strong intellectual ties, religious convictions, and information networks that exceeded their countries'

[35] In the words of Sarah Hutton, Cambridge Platonists advocated theological optimism and clashed with "the harsh predestinarian theology of Calvinism." They emphasized freedom of the will and insisted on the "importance of reason in religion." Furthermore, "by widening the traditional Platonic defense of the immortality of the soul to a defense of spirit in general, they sought to combat what they saw as atheistic tendencies in contemporary thought." They saw determinism and atheistic materialism in the works of Hobbes and Spinoza—and even to some extent in that of Descartes. The Platonists' opposition to materialist philosophies equaled their rejection of Calvinist predestination. See Sarah Hutton, "Introduction to the Renaissance and the seventeenth century," in *Platonism and the English Imagination*, 73. For Henry More specifically, see Robert Crocker, "Henry More: a biographical essay," in *Of Mysticism and mechanism: Tercentenary studies of Henry More (1614–1687)*, ed. Sarah Hutton (Dordrecht, NL: Kluwer Academic Publishers, 1990), 4.

[36] Sir William Temple of Shene, *Observations upon the United Provinces of the Netherlands* (London: n.p., [1673] second ed. 1680), 242.

intertwined dynasties.[37] To the English, the Dutch Republic was a safe haven away from political persecution and religious troubles at home. When John Locke was accused of plotting against the English Crown, he fled to Rotterdam in 1683 and later moved to Amsterdam. His political exile lasted until February 1689 and had a profound impact on his life and creativity. Among the Dutch, Philip van Limborch actively positioned himself as an ambassador for the Remonstrant cause abroad. He not only emerged as the leading representative of the Remonstrant Brotherhood of his time, but he also sought "to convey the truth" to English theologians and clergy about the Remonstrant ideological standpoint and beliefs.[38] He and his Remonstrant circle were invested in advancing "a rational, anti-fanatical, anti-tyrannical view of Christianity such as that of the English latitudinarians and neo-Platonists, the Socinians, and the French Protestants."[39] The Revocation of the Edict of Nantes officially closed France to reformed religion in 1685. As a result, the United Provinces were inundated by French Protestant preachers. Faced with unemployment at the pulpit, religious refugees and exiles, such as Pierre Bayle, Jean Le Clerc, Pierre Coste, and Pierre Desmaizeaux, channeled their preaching into journalism and correspondence.[40]

Limborch's circle of correspondents comprised his compatriots and fellow theologians—Christiaan Hartsoeker, Jacob Johannes Batelier, Nicolaas Borremans—as well as the physician Lambert van Velthuysen, the lawyer Gerard Noodt, the English Earl of Shaftesbury, and the philosopher John Locke, who authored *Essay on Toleration* in 1667. His vast network stretched east and west across fluid national boundaries. It included the German theologians Daniel Wülfer and Philippus Hieronimus Andreae, and the French theologians Louis le Blanc and Pierre Allix. These men exchanged their thoughts and ideas about theological and philosophical matters and frequently sent each other their newly printed publications. But they also informed each other about

[37] Some of the pioneering works include Rosalie L. Colie, *Light and Enlightenment: A Study of the Cambridge Platonists and the Dutch Arminians* (Cambridge, UK: Cambridge University Press, 1957); J. S. Bromley and E. H. Kossmann, eds., *Britain and The Netherlands. Papers Delivered to the Oxford-Netherlands Historical Conference* (London: Chatto and Windus, 1960); and C. L. Thijssen-Schoute, Uit de Republiek der Letteren ('s-Gravenhage, NL: Martinus Nijhoff, 1967). More recent works include Luisa Simonutti, "Religion, Philosophy, and Science: John Locke and Limborch's Circle in Amsterdam," in *Everything Connects: In Conference with Richard H. Popkin* (Leiden, NL: Brill, 1999), 295–324; Jori Zijlmans, *Vriendenkringen in de zeventiende eeuw: Verenigingsvormen van het informele culturele leven te Rotterdam* (The Hague, NL: Sdu Uitgevers,1999); and Lisa Jardine, *Going Dutch: How England Plundered Holland's Glory* (New York: HarperCollins, 2008).

[38] Pieter Jacobus Barnouw, *Philippus van Limborch* (The Hague, NL: Mouton & Co., 1963), 24.

[39] Luisa Simonutti, "Religion, Philosophy, and Science: John Locke and Limborch's Circle in Amsterdam," in *Everything Connects: In Conference with Richard H. Popkin* (Leiden, NL: Brill, 1999), 308.

[40] Rosalie Colie, "John Locke in the Republic of Letters," in *Britain and the Netherlands*, eds. J. S. Bromley and E. H. Kossmann (London: Chatto & Windus, 1960), 120–121.

their personal lives.[41] Both Locke and Limborch actively debated the relationship between political and ecclesiastical power, and argued passionately and insistently for religious toleration.[42] One of the reasons that led Limborch to establish connections with his English colleagues was to inform them specifically of the Synod of Dordrecht and its very real consequences. Some religious groups had accused Remonstrants of slandering the evangelical and reformed religions.[43] The Calvinists' attack on the Remonstrant faith in 1618–1619 culminated in their expulsion from the Dutch Republic. This relatively small religious minority was not allowed to repatriate until the 1680s. With the goal of setting the record straight, Limborch sent Henry More four copies of a published compilation of letters (from both Dutch and English theologians and thinkers). Limborch had originally co-edited them with Hartsoeker's father Christiaan in a volume entitled *Praestantium ac Eruditorum Virorum Epistolae Ecclesiasticae et Theologicae varii argumenti: Inter quas eminent eae, quae à Iac. Arminio, Conr. Vorstio, Sim. Episcopio, Hug. Grotio, Caspar Barlaeo conscriptae sunt.* While one copy was destined for More, the other three were meant to reach the hands of Ralph Cudworth, Oliver Doiley, and Henry Jenkes.[44]

Limborch's and Christiaan Hartsoeker's intellectual circle discussed, wrote, and lectured on a wide variety of subjects, ranging from theology to natural philosophy. Although Limborch was not an outspoken Cartesian, he was well versed in Descartes's natural philosophy. In a letter to Hartsoeker's father, Limborch disclosed how he let his students read at home the then-unorthodox works by Gassendi and Descartes: "Every day I myself will examine them, explain ambiguities, and expand on their doubts," so that his students "will all perceive the thoughts of the ancients in philosophy and will be able to read all Philosophers' books, even Descartes', without an instructor."[45] The young Hartsoeker himself reported to have been a student of Limborch for a brief period of time in Amsterdam. It seems highly plausible that he was instructed in, or at least aware of, the writings of various orthodox and less orthodox contemporary

[41] Barnouw, *Philippus van Limborch*, 24–26.

[42] Simonutti, "Religion, Philosophy, and Science," 307.

[43] *Nederlandse Hervormde Kerk. Synoden, A proclamation giuen by the discreet lords and states, against the slanders laid vpon the euangelicall and reformed religion, by the Arminians and separatists containing all the points, accusations, declarations and confessions, taken out of the last prouinciall synode holden at Arnhem, the 15. day of September last past. 1618. Together with the seuerall examinations and confessions (at Vtrecht and the Hage) of one Leydenberg, pentioner of Leyden, and Taurinus; with their sodaine and fearefull ends* ([London]: Printed according to the Dutch originals, at London, by G[eorge] E[ld] for Th[omas]: Th[orpe]: and Richard Chambers, and are to be solde at the signe of the blacke Beare in Paules Church-yard [at the shop of Edward Blount], 1618).

[44] Barnouw, *Philippus van Limborch*, 15–16.

[45] Philip van Limborch to Christiaan Hartsoeker, Universiteitsbibliotheek Amsterdam, Special Collections, III D 17, fol. 95r (undated). See also Barnouw, *Philippus van Limborch*, 26.

thinkers. Consequently, it would not be a far stretch to suggest that Hartsoeker's Arminian upbringing had "made him sympathetic to Platonism," since "his father Christiaan had belonged to a circle of Arminian theologians [who] corresponded with Henry More, Ralph Cudworth, and other Cambridge Platonists during the 1660s and 1670s."[46] What seemed to trouble these circles was "the apparent determinism of Cartesian physics, which seemed to disallow both human free will and divine intervention in the world."[47] These men had debated Descartes's matter in motion a generation before Hartsoeker, supplying him with formidable counterarguments.

4. Henry More's Extension

Henry More was one of the key Cambridge Platonists whose published letters to René Descartes highlighted the inconsistencies in the Cartesian system and inspired Hartsoeker to rethink the principles of physics. More was a moderate Anglican divine and theologian known for his revival of a Platonist interpretation of Cartesian physics and his opposition to dogmatic Calvinism. More kept abreast of developments in the new science: He was an early member of the Royal Society (elected on May 25, 1664[48]) and even constructed experiments himself. However, "More viewed reason or intellect as the first mover in philosophical inquiry, with experiments playing a useful supporting role to rational arguments."[49] Based on More's early interest in Descartes's work, some of his fellow Cambridge Platonists—most notably Samuel Hartlib and Ralph Cudworth—encouraged him to write to Descartes.[50] Henry More's reactions to Descartes's ideas were by no means uniform. Over time, his earlier ardent admiration for the French philosopher gave way to a more nuanced critical reassessment of Cartesian philosophy and how it applied to his own philosophical and religious concerns. But, Alan Gabbey argues, More's basic philosophical stance

[46] Alice Stroup, "Nicolas Hartsoeker: Savant, Secret Agent, Man of Conscience?" (paper presented at the Annual Meeting of the History of Science Society, Santa Fe, NM, Nov 12, 1993), 11–12. See also Colie, *Light and Enlightenment*, 7 ff.

[47] Stroup, "Nicolas Hartsoeker: Savant, Secret Agent, Man of Conscience?" 11; Colie, *Light and Enlightenment*, xi.

[48] The Royal Society of London, *List of Fellows of the Royal Society, 1660–2007*, A–J (July 2007), accessed on Jan. 3, 2012 at http://royalsociety.org/uploadedFiles/Royal_Society_Content/about-us/fellowship/Fellows1660-2007.pdf

[49] Crocker, "Henry More: a biographical essay," 8.

[50] Charles Webster, "Henry More and Descartes: Some New Sources," *The British Journal for the History of Science* 4 (1969): 364. Webster cites a letter from Henry More to Samuel Hartlib from Nov. 27, 1648.

on Cartesian ideas did not change significantly after 1650.[51] By this time, "More had acquired a comprehensive and detailed knowledge of most of Descartes's contributions to philosophy, and had developed an all-embracing and lasting interest in their implications for what More considered were the important theological and philosophical issues of the time."[52] The correspondence between the two men laid bare some of the central criticisms of Descartes's philosophy.[53] The physical universe expressed a rational and moral order that More saw confirmed in Descartes's rational and systematic explanation of the universe.[54] However, "Descartes' matter in motion seemed unable to suggest anything else than the active informing presence of some poorly articulated spiritual substance."[55] To More, the world's "inherent order, providential design and rationality referred not simply to the goodness of God, but to a Neoplatonic spiritual hierarchy of active and intelligent spirits informing and upholding the 'exterior' experienced world of the senses."[56] As such, he argued that there must be a divine presence or spirit that infuses all matter.

Nonetheless, there were some crucial differences between More's and Descartes's conceptions of creation. The English thinker addressed them in his 1648–1649 correspondence with Descartes. At the time, he was primarily concerned with understanding the Frenchman's ideas more fully.[57] Unlike Descartes, Henry More counted God among extended things.[58] Using this idea as foundational, More explained that "human souls, all good and bad spirits are corporeal, and consequently have a real sensation, that is, a sensation that comes

[51] Alan Gabbey, "Philosophia Cartesiana Triumphata: Henry More (1646–1671)," in *Problems of Cartesianism*, eds. Thomas M. Lennon et al. (Kingston and Montréal: McGill-Queen's University Press, 1984), 194, 214.

[52] Ibid., 191.

[53] Ibid., 192–193. For the interested reader, Gabbey summarizes 11 points of disagreement between More and Descartes in concise but not exhaustive fashion, as he admits himself.

[54] Robert Crocker, introduction to *The Life of Henry More*, by Richard Ward, eds. Sarah Hutton et al. (Dordrecht, NL: Kluwer Academic Publishers, 2000), xvii.

[55] Crocker, introduction, xviii.

[56] Ibid.

[57] Henry More to Samuel Hartlib, Aug. 27, 1649: "I am glad I shall hear from him before he takes this long journey [to Sweden]. For this letter answering is more consyderable to me then the answering of any I wrote yett to him. And the two next I intend to write will be muchwhat of the same importance for my designe, viz. the thorough understanding his philosophy, that I may be self a safe and able judge of it." And also see More to Hartlib, Sept. 24, 1649: "And now I begin somewhat eagerly to expect an answer from Des Cartes of my last letter [July 23, 1649] to him. For I can settle my self to nothing to any purpose till I have thoroughly perused his Philosophy, nor can I go on in the perusall thereof till such time as I hear from him. . . ." Both letters cited in Gabbey, "Philosophia Cartesiana Triumphata," 196.

[58] Henry More to René Descartes, Cambridge, Dec. 11, 1648, in Henry More, *A Collection of Several Philosophical Writings*, 2 vols., ed. René Wellek, vol. I (New York: Garland Publishing, Inc., 1978), *Henrici Mori Epistolæ Quatuor ad Renatum Des-Cartes*, 62.

from the body which they inhabit."[59] Descartes disagreed. In a letter to More, he supposed that a human soul separated from the body felt nothing whatsoever. More asked Descartes if a philosopher could not accept an incorporeal substance in nature that endowed matter with properties such as motion, figure, and general structure. Couldn't there be an incorporeal substance that resided in the body and governed how that body moved, divided, united, dissipated, and organized its parts?[60] Descartes responded that all a philosopher could grasp about the nature of immaterial substances and the soul were conjectures, and could not be accounted for by mechanistic explanations.[61] Regarding material substances, however, Descartes wrote that he had "found nothing on the[ir] nature...for which [he] could not very easily give a mechanical reason."[62] Essentially, motion was the only glue that bound matter particles together.[63] Descartes's God had infused matter with motion in the moment of the world's creation, but did not actively manage matter at every instance. God merely conserved what motion he had granted in the beginning. Descartes avoided discussing this question for fear of "seeming to favor those who liken God to the soul of the world that imbues matter."[64] Thus ended the direct correspondence between More and Descartes. More's *Collection of Several Philosophical Writings*, published in 1662, contained the correspondence. So by 1678, the year when Hartsoeker was a fourth-year student at Leiden University and about to travel to Paris for the first time, all of Henry More's works were freely available. Although it is impossible to pinpoint the precise moment of Hartsoeker's break with the Cartesian system of nature, it is safe to say he began taking More's Platonist ideas seriously around 1696.

That was the year when Hartsoeker's growing unease with Cartesian physics became evident and when he cited Henry More's correspondence with Descartes in an article in *Le Journal des Sçavans*. Here, Hartsoeker responded to a Cartesian professor of mathematics and philosophy, La Montre, who attacked Hartsoeker's newly published treatise *Principes de Physique*. Building on More's objections to Descartes, Hartsoeker defended his own cosmological views.[65] For one, he (like More) argued that Descartes had not fully explained the first principles that underpinned matter in motion. They both sensed that Descartes had

[59] More to Descartes, Cambridge, July 23, 1649, *A Collection of Several Philosophical Writings*, 86.

[60] More to Descartes, Cambridge, Mar. 5, 1649, *A Collection of Several Philosophical Writings*, 80.

[61] Descartes to More, Endegeest, Aug. 1649, *A Collection of Several Philosophical Writings*, 103–104. This letter manuscript was found among Descartes's papers in which he had begun crafting a response to More's two preceding letters to him. Descartes's response was never sent to More, but appeared in More's published writings in 1662 and concludes their correspondence.

[62] Descartes to More, Egmond, May 17, 1649, *A Collection of Several Philosophical Writings*, 84.

[63] Ibid.

[64] Descartes to More, Endegeest, Aug. 1649, *A Collection of Several Philosophical Writings*, 104.

[65] Stroup, "Nicolas Hartsoeker: Savant, Secret Agent, Man of Conscience?" 11.

told only half of the story: For example, the French philosopher gave a rather sound account of matter particles but not of the subtle matter in which they moved.[66] According to Hartsoeker, Descartes supposed that this subtle matter was neither fluid nor hard, yet had infinite extension. Both Hartsoeker and More conceived of extension as something that communicated essence and movement. Both believed it to be corporeal and real, though, in More's words, "more general than the body."[67]

The Dutchman believed that he could solve the objection that More raised against Descartes's system by reducing everything to two essential properties, liquidity and hardness. In Hartsoeker's system of the world, these two qualities underpinned all fundamental principles.[68] Matter particles were absolutely hard and infinite in number, while the liquid first element was perfectly fluid and imparted motion to the atoms. Hartsoeker trumpeted his invention of the fluid first substance with the following words: "[I]t has all the advantages of Gassendi's void and Descartes's subtle matter, and none of their inconveniences."[69] Hartsoeker did not elaborate further. At this point, he did not ascribe any kind of intelligence or consciousness to the first substance of his physical system. Ironically and perhaps prematurely, Hartsoeker added that he did not want to entangle himself in "further metaphysical subtleties which have nothing to do with the work of a Physicist. . . ."[70] That is, Hartsoeker did not argue for a spirit of nature or any additional metaphysical forces in his system. Instead, he presented a reasoned account of where Descartes's system withstood and where it faltered under the pressure of outside arguments. Nonetheless a philosophical change had begun to take shape. For one, Hartsoeker then went on to save the phenomena with his own system of nature, which contained adaptations in the language of More's correspondence with Descartes.

5. Ralph Cudworth's "Plastick Nature"

The arguments Hartsoeker adopted from Ralph Cudworth had their origins in an active philosophical school with whose ideas he clearly found useful to think. Cudworth's revival of "Plastick Nature" probably entered Hartsoeker's

[66] Nicolas Hartsoeker, "Des Elemens du Corps Naturel, et des qualitez qu'ils doivent avoir," *Journal des Sçavans*, July 16, 1696: 326–327.

[67] Ibid., 327. Cf. More to Descartes, Dec. 11, 1648, in More, *A Collection of Several Philosophical Writings*, vol. I, 62. See also More to Descartes in *Descartes in Correspondance [de René Descartes] avec Arnaud et Morus*, ed. Geneviève Lewis (Paris: J. Vrin, 1953), 99.

[68] Hartsoeker, "Des Elemens du Corps Naturel, et des qualitez qu'ils doivent avoir," 329.

[69] Ibid.

[70] Ibid.

philosophy after 1703.[71] Cudworth rejected two versions of materialism that he considered atheistic: atomism and hylozoism. The former presupposed that only matter existed, while the latter accepted self-organizing life in matter. Cudworth wrote especially against the atomists, whom he saw as atheists (namely, Hobbes, Spinoza, and their ilk). In 1678, his major work, *The True Intellectual System of the Universe*, lumbered off the presses. But the work was not widely read, as it was not translated until 1733. In 1703, it resurfaced on the continent: Jean Le Clerc, encouraged by his friend John Locke, published a series of extracts from Cudworth's book in French in his *Bibliothèque Choisie*.[72] Cudworth's excerpts on plastic nature "pleased all but the most stubborn Cartesians and Spinozists." However, the vitriol between Le Clerc and Pierre Bayle that exploded upon their publication famously rocked the lettered republic.[73] In this work, Cudworth developed the famous doctrine of plastic nature, which he derived from Plotinus's concept of a vegetative soul. (Incidentally, Hartsoeker also kept Plotinus's *Opera Philosophica Omnia* from 1580 and Plato's complete works in his library.[74])

Essentially, "Plastick Nature" was "the explanation for the laws of nature" without invoking God's direct operation in the world. It was God's instrument, unaware and unconscious of its own nature but expressing God's order in material things.[75] To Cudworth, it was not an occult cause; rather, it was "a manifestation of 'mental causality' in the world."[76] Cudworth believed that God, or rather the divine principle of love and goodness, expressed itself in all living things as "the Plastick Nature."[77] Being alive entailed having a kind of "organic unity," a phrase used by Leslie Amour to describe how Cudworth's conception of free will played out in tandem with Plastick Nature and the metaphor of an organic body. In other words, the human body, as that of other living beings, had "a coordinating power" that could direct action in order to express "the character of its natural order."[78] At its basis, "freedom requires a kind of organic unity. It is not

[71] Although Hartsoeker did not explicitly mention Cudworth by name as the originator of his ideas for intelligent forces of nature, he mentioned "some authors [who] have had recourse to a plastic Nature" in his *Seconde Partie de la Suite des Conjectures Physiques* (Amsterdam: Nicolas Viollet, 1712), 20. Nonetheless, Fontenelle saw him borrowing Cudworth's idea of "Plastick Nature." See, Bernard le Bovier de Fontenelle, "Eloge de Monsieur Hartsoeker," *Éloges des Académiciens de l'Académie Royale des Sciences, morts depuis l'an 1699*, Tome II (Paris: Libraires Associés, 1766), 242.

[72] Colie, *Light and Enlightenment*, 117.

[73] Colie, "John Locke in the Republic of Letters," 127–128.

[74] Nicolas Hartsoeker, *Bibliotheca Hartsoekeriana* (The Hague, NL: Johannes Swart, 1727), 119.

[75] Colie, *Light and Enlightenment*, 128.

[76] Leslie Amour, "Trinity, Community and Love: Cudworth's Platonism and the Idea of God," in *Platonism at the Origins of Modernity: Studies on Platonism and Early Modern Philosophy*, eds. Douglas Hedley and Sarah Hutton (Dordrecht, NL: Kluwer Academic Publishers, 2008), 118.

[77] Ibid., 120.

[78] Ibid., 115.

just accident that we are put together in such a way that our nature requires a unifying—and free—force."[79] Rather than speak of "organic unity," Cudworth discussed the necessity for an internal ruler in the body: "[S]ome one hegemonical" in us "which comprehending all the other powers, energies, and capacities of our soul... determineth, not only actions, but also the whole passive capability of our nature one way or another, either for the better or the worse." Such a hegemonic principle would have to act like glue and govern a structure as complex as a living human or animal machine. Without an internal hegemon,

> [w]e should be like to a disjointed machine or automaton all whose wheels are not well set together; which therefore will be either at a stand continually, or else go on very slowly heavily and cumbersomely. It could never carry on evenly any steady designs, nor manage itself orderly and agreeably in undertaking, but would be altogether a thing inapt for action.[80]

As Cudworth wrote, "God Almighty could not make such a rational creature as this is, all [*sic*] whose joints, springs, and wheels of motion were necessarily tied together, which had no self-power, no hegemonic or ruling principle, nothing to knit and [unite] the multifarious parts of the machine into one."[81] In these last two quotations from Cudworth, the metaphor of the machine figures as both an invaluable and yet troubling mental aid: invaluable, because it so aptly and elegantly explains most living beings; and troubling, because it nonetheless fails to explain the problem of the prime mover by virtue of being exactly that, a machine. Crucially, it highlights the evolving problems thinkers were left with when confronted with the Cartesian machine and the question of whether or not animals had souls.

Hartsoeker, however, amended Cudworth's Plastick Nature by attributing conscience or knowledge ("*connoissance*") to it. The fact that Cudworth's formative principle acted unconsciously abetted atheism in the eyes of his contemporary critics.[82] Meanwhile, Hartsoeker's tweaking of the formative principle involved endowing it with Intelligence, as we saw in the crayfish experiments.[83]

[79] Ibid.

[80] Ralph Cudworth, *A Treatise of Freewill* [first ed. and publ. from Cudworth's ms. c. 1654 by John Allen] (London: John W. Parker, 1838), 58–60.

[81] Cudworth cited in Amour, "Trinity, Community and Love," 115.

[82] Roger, *The Life Sciences in Eighteenth-century French Thought*, 346.

[83] Hartsoeker, *Seconde Partie de la Suite des Conjectures Physiques* (Amsterdam: Nicolas Viollet, 1712), 20. See also Fontenelle, "Éloge de Monsieur Hartsoeker," 242.

6. Intelligent Forces and the World Soul

In 1706, Hartsoeker wrote that nature's ways puzzled the human mind. In physics, "one is often obliged to admit probabilities in lieu of demonstrations." Because atoms could not be perceived through the senses, they had to be inferred through their effects.[84] Hartsoeker seemed hopeful that "a small number of simple and uniform laws" governed the infinite effects of nature. "[I]n my system," he wrote, "everything was the eternal and direct ['*immediate*'] work of God" and "was composed of immutable and indivisible particles."[85] He likened his first element to pure fire, whose nature we could not know, yet whose being was "real" and could "push the atoms or little bodies floating in it, and keep them from colliding with one another."[86] The second element, or collectively matter particles, he described as small solid masses that were simple, homogenous, and perfectly hard, but incapable of any movement by themselves.[87]

In subsequent years, further traces of Neoplatonist ideas appeared in Hartsoeker's ever more nuanced system. Now intelligence and consciousness began to play a role in his proposed first element. In a 1710 letter to Leibniz, Hartsoeker described his first element as "pure fire and the soul of the universe."[88] This first substance was endowed with intelligence, by which it was able to arrange matter without end. This Intelligence was the source of atomic motion and of many other subsidiary Intelligences, or souls.[89] Some of these subordinate souls had the power to maintain, for example, an entire planetary system, while others worked "continuously on the creation of animals and plants."[90] Hartsoeker's first element seemed a bottomless pit of Intelligences, each of which had its own special function in the guidance and conservation of matter. Consequently, Hartsoeker claimed that the life of the world's soul was fueled by these thoughts or intelligent forces. Conversely, "... the total extinction of thought would be the soul's [and thus, the universe's] actual death...."[91] To Fontenelle, Hartsoeker's "prodigious number of Intelligences spread throughout" seemed

[84] Hartsoeker, *Conjectures Physiques* (Amsterdam: Henri Desbordes, 1706), 1–2.

[85] Hartsoeker, "Avertissement" to *Conjectures Physiques*, unpaginated.

[86] Hartsoeker, *Suite des Éclaircissemens sur les Conjectures Physiques* (Amsterdam: Nicolas Viollet, 1712), 2.

[87] Nicolas Hartsoeker to Gottfried Wilhelm Leibniz, Düsseldorf, Dec. 30, 1710, in C. I. Gerhardt, *Briefwechsel zwischen Leibniz und Hartsoeker*, III, 514.

[88] Hartsoeker to Leibniz, Düsseldorf, Aug. 22, 1710, Gottfried Wilhelm Leibniz Bibliothek, Hannover, DE: LBr. 317. Also see Gerhardt, III, 502.

[89] Hartsoeker used these two terms interchangeably.

[90] Hartsoeker to Leibniz, Düsseldorf, Aug. 22, 1710, Gottfried Wilhelm Leibniz Bibliothek, LBr. 317. Also see Gerhardt, III, 502–503.

[91] Hartsoeker to Leibniz, Düsseldorf, Dec. 30, 1710, LBr. 317. Also see Gerhardt, III, 516.

a revived notion from ancient philosophy. Fontenelle decried these intelligent forces, which he had "believed [to be] abolished forever." Evidently, they had not been "banned enough" since they managed to resurge in "modern" philosophy, as Hartsoeker's invocation of Intelligences showed.[92]

Hartsoeker did not see it Fontenelle's way. At the time, Cudworth and Malebranche had proposed two probable but problematic explanations of the relationship between body and soul. While Hartsoeker considered both, he was concerned with how each adjudicated God's role, causality (broadly defined), free will, and animal soul. Hartsoeker read the two authors in the following way. In *The Search After Truth*, Malebranche explained voluntary movement in animals and humans with occasional causes. As such, all of man's actions were caused directly by God. Similarly, Cudworth's Plastick Nature, which worked without conscious choice and knowledge, was but another manifestation of God's will over all of creation.[93] In any case, both propositions eliminated free will.

In response to the constraints of Cartesian, Malebranchian, and even Neoplatonist theories, Hartsoeker launched the spiritual aspects of his system in the *Seconde Partie de la Suite des Conjectures Physiques*, published in 1712, the year of Réaumur's crayfish experiments. His conception of active souls boiled down to the following: Each living creature—animal or human—had two souls, a vegetative soul that was subordinate to the rational soul inhabiting the same body. Whereas the vegetative soul was in charge of involuntary bodily functions, the rational soul controlled voluntary actions. The vegetative soul, or "animal spirits," had the task to "strictly obey" the "orders" of the rational soul, which ultimately governed all operations between it and the body, and between it and God. Hartsoeker explained his thinking with a metaphor: To make a machine work, one would need

> either a weight or some spring mechanism that controls the entire machine, or some artisan, and most often the same Artisan who put the machine together. And seeing that there is neither a weight nor a spring in our Machine to make it run, one must have recourse to a Being or some Artisan who knows the entire construction of this Machine perfectly.... And so, why shouldn't I have recourse to an intelligence which resides in me and who knows all the organs inside and out, and acts upon the orders she receives from my Soul to the best of her abilities?[94]

[92] Fontenelle, "Éloge de Monsieur Hartsoeker," 247.

[93] Hartsoeker, *Seconde Partie de la Suite des Conjectures Physiques*, 19–20.

[94] Ibid., 15–16.

Hartsoeker speculated that each of these souls could very well be "a part of my first Element, which I have regarded as the Soul of the Universe, and which . . . has received the faculty of feeling, consciousness, and thought" from God.[95]

Our Dutchman believed his ideas satisfied where Father Malebranche's occasionalism and Ralph Cudworth's Plastick Nature disappointed. Crucially, they proposed systems that omitted free will altogether. This is something Hartsoeker repeatedly discussed in his treatises published between 1712 and c. 1730. He believed his intermediary system of intelligent and conscious spirits successfully addressed the inconsistencies in the Malebranchian and Cudworthian systems. If, on the one hand, occasionalism was true, then God did "nothing but continually obey man's will." Occasionalism essentially meant that God alone brought about all phenomena of the body and the mind in accordance with a few simple laws of nature. If, on the other hand, a plastic nature existed, then it worked "continuously with art and method towards a certain goal intended by God, but without choice, discernment ['*discernement*'] and knowledge of the technique ['*artifice*'] of its own Works." Here, once again, "God [did] everything himself."[96] Clearly, Hartsoeker was frustrated with a metaphysics that posited God as an active force present in, and responsible for, every action of every living being. Every phenomenon would then become a miracle. The theory of Plastick Nature seemed too close in spirit to Malebranche's occasionalism, and also suggested a passive world in which God ultimately did everything himself. At the same time, Hartsoeker had clearly rejected Descartes's vision in which God watched the world unfold according to a set of preordained mechanical laws. To some extent, Hartsoeker's invention of Intelligences solved this philosophical impasse. After all, Intelligences or souls were the intermediary forces between God and matter, and they preserved freedom of will. They were responsible for the living being from beginning to end, and learned how to adapt its body to the external environment.

What was God's role then? In Hartsoeker's universal scheme, God was everywhere and yet not directly present. That is, God acted through the first element that arranged the atoms to form sensible bodies. In sum, "the Universe is like an animal, full of life and intelligence, because I conceive the first element to be endowed with life and intelligence under the direction of God, and to move the bodies incapable of moving themselves."[97] In his scheme, every living thing had an intelligent soul. To be alive was to think. As a result of this doctrine, the Jesuits of Trévoux accused him of ascribing divinity to the universe itself. Leibniz was

[95] Ibid., 147–148.

[96] Ibid., 19–20.

[97] Hartsoeker, *Suite des Eclaircissemens sur les Conjectures Physiques* (Amsterdam: Nicolas Viollet, 1712), 67.

also shocked at his ideas. He reproved Hartsoeker: If the universe is like an animal, then one may take it to mean that God is the soul of this animal, instead of being the *Intelligentia supramundana*.[98] In this respect, Hartsoeker's formulation seems a fusion of cosmic vitalism and immanent vitalism, as defined by Kevin Chang. The former "presumed a principle of life that originated in the heavens, pervaded the cosmos and gave both life and form to corporeal beings. This principle of life was often identified as astral bodies, the universal spirit or *anima mundi*." The latter, in contrast, "postulated a vital principle that was embodied in, and immanent to, the most fundamental unit of matter, the monad or *semina*, for example. This vital principle was the form of an organism, thus its formal cause. It was also its efficient cause, as it directed the developmental course of the organism. It was even the final cause of the organism, considering that it incorporated in itself the end of the living organism."[99]

This schema becomes more evident in Hartsoeker's letter to Jean Le Clerc that was published posthumously in 1730 in a compilation of writings entitled *Cours de Physique*. There, Hartsoeker gave more concrete form to the Intelligence he had introduced after the crayfish experiment. This very "same Intelligence" had "invisibly built & constructed this animal in the body or, better said, in one of the testicles of the male, which had fathered it, ... with art and science, in accordance with the power she received to do this from a superior Intelligence." Like an internal engine, the Intelligence took care of this animal as much as she could over the course of her life. "It seems without question to me that there are subaltern Intelligences to God that have a portion of matter, which they may arrange freely, under their own direction," Hartsoeker hypothesized. After all, these Intelligences made mistakes and "fail[ed] in their creations." "We have ourselves as examples," he wrote, "for who would like to maintain that God Himself moves our limbs, & that the Intelligence, which lives in us indisputably, has no other role but to will this; and that God Himself makes the blood & the humors in our body circulate continuously by means of the systole & the diastole of the heart, & by the peristaltic movement of veins, through which the blood & the humors flow; or that all this is done by the laws of motion?"[100] This late text raises a spate of questions. Did Hartsoeker not think it problematic that animals had thinking souls or Intelligences that guided them from existence onward and left them at death? And why should there be two separate souls, a vegetative and a rational

[98] Leibniz to Hartsoeker, Hannover, Feb. 6, 1711, GWLB, LBr. 317 (or Gerhardt, III, 520).

[99] Ku-Ming (Kevin) Chang, "From Vitalistic Cosmos to Materialistic World: The Lineage of Johann Joachim Becher and Georg Ernst Stahl and the Shift of Early Modern Chymical Cosmology," in *Chymists and Chymistry: Studies in the History of Alchemy and Early Modern Chemistry* (Sagamore Beach, MA: Science History Publications, 2007), 221.

[100] Hartsoeker, "Lettre a Monsieur Le Clerc sur quelques endroits de la Philosophie Newtonienne," *Cours de Physique*, 1730, 8.

soul, in an animal rather than one? And what about the problematic fact that his Intelligences had finite "lives," as he seemed to imply here? In his contrivance of Intelligences, was he himself not resorting to occult causes, for which he frog-marched Newton and others with vainglorious passion?[101]

More confusing still is Hartsoeker's ambiguous use of Intelligence and soul as synonymous. Evidently, the question of souls or Intelligences occupied him at least throughout the last 15 years of his life. Toward the end, Hartsoeker still puzzled over the origin and nature of this Intelligence, with which everything in the universe was imbued:

> ...I resolved finally to conjecture that there is only one soul in us that does everything. Moreover, I conjectured that this soul might well be nothing other than a portion of the soul of the universe, which I called in my works on physics the first element or perfectly fluid substance, and which, having extension like matter—although otherwise essentially different from it—can impel bodies and be impelled by them, give them movement and receive movement from them; and I resolved all the more readily to make this conjecture in that I freed myself thereby from the great difficulty that men have always had in conceiving how communication can occur between soul and body. In this way, I conjectured that the soul is spread out through the whole body....[102]

In characteristic fashion, Hartsoeker let his conjectures explain how the soul thinks and behaves. He believed his thinking to be free from the artificial and unproductive construct of the dichotomy between body and soul. While the soul was "a wish and gift of [the infinite and all-powerful] God," whose nature was out of his reach, conjectures could approximate how communication between mind and body occurred.[103]

7. Conclusion

Hartsoeker's notion of Intelligences in explaining regenerative phenomena, as those in crayfish, had an afterlife. It found resonance in later experiments

[101] Christiane Berkvens-Stevelinck, "Nicolas Hartsoeker contre Isaac Newton ou pourquoi les planètes se meuvent-elles?" *Lias* 2 (1975): 313–328.

[102] Hartsoeker, "Remarques sur une These de Physique...," *Recueil de plusieurs pieces de physique*, 195–196. Here I follow Robert Ellrich's translation of this passage in Roger, *The Life Sciences in Eighteenth-century French Thought*, 349.

[103] Hartsoeker, "Remarques sur une These de Physique...," *Recueil de plusieurs pieces de physique*, 197–198.

conducted by Charles Bonnet, Abraham Trembley, and Réaumur on the regenerative faculties of worms and polyps in the early 1740s. The question of animal soul continued to animate the investigations of these naturalists and animal experimenters. During the sectioning of worms and polyps, Bonnet witnessed that they not only remained alive, but also reproduced their amputated parts. Like Hartsoeker, Bonnet remained skeptical of interpretative frameworks such as Descartes's soulless animal automata and Malebranche's passive *emboîtement* theory (where germs were encased one within the other from the beginning of life and time). However, he was definitely intrigued by Jean-Louis Calandrini's system of germs. Calandrini's theory posited that germs with soul-like properties caused the reproductions in polyps.[104] In some respects, his hypothesis for the regrowth of limbs in crayfish was similar to that proposed sketchily by Réaumur. Bonnet found Calandrini's explanation so convincing because it provided a metaphysical basis for the polyp's organizational complexity and functional unity. Trembley's experiment of cutting the polyp in sections lengthwise should have destroyed the polyp's "Animal Oeconomy." Instead, Bonnet, like Hartsoeker with the crayfish, believed the polyp's regenerative behavior suggested the possibility of a soul, similar to that in man.[105] In a 1742 letter to mathematician Gabriel Cramer, Charles Bonnet asked if "[t]he partisans of the system of Mr Hartsoeker's *Intelligences rectrices* would not think themselves well supported by a similar experiment [on the polyp]?"[106] Bonnet would only read Leibniz's *Theodicy* in 1747, after reflecting on Trembley's, Réaumur's, and his own experiments on aphids.[107] Although Hartsoeker did not appear troubled by the consequent metaphysical impasse of animal soul that his Intelligences inevitably gave rise to, later experimenters like Bonnet definitely were. After all, Intelligences begged the question of whether such forces truly existed and had a demonstrable causal explanation.

Hartsoeker's gradual modifications to his Cartesian-inspired atomist system of nature show that he, like others before and after him, was struggling

[104] Virginia Parker Dawson, *Nature's Enigma: The Problem of the Polyp in the Letters of Bonnet, Trembley and Réaumur* (Philadelphia: American Philosophical Society, 1987), 160.

[105] Bonnet's "Animal Oeconomy" cited in Dawson, *Nature's Enigma*, 161.

[106] Bonnet cited in Dawson, *Nature's Enigma*, 163 and n163: "Je souhaiterois bien de scavoir, mais je crains, Monsieur, d'abuser de vôtre complaisance, si l'expérience que je fais de couper à un même par la tête et la queuë à mesure qu'elles lui reviennent vous paroît aussi curieuse qu'elle me le semble. Elle peut fournir au moins des eclaircissemens à plusieurs questions interessantes dont j'ai cotté quelques unes dans ma Lettre à M. de R. de 23 Juin dernier. Je fais aujourdhuy la 7e coup. Les partisans du systeme de Intelligences rectrices, du M. Hartsoêker, ne seroient-ils pas cru[?] bien appuyés par une semblable experience?" Bonnet to Cramer, Aug. 17, 1742, ms. suppl. 384, Bibliothèque Publique et Universitaire de Genève. Based on Dawson's citation of the original letter (see above), I have somewhat changed her English translation of the last sentence.

[107] Dawson, *Nature's Enigma*, 177.

with the problems posed by Descartes's philosophy. After all, the battle for a coherent explanatory model of nature in a post-Cartesian world was at hand before Réaumur's crayfish experiments in 1712. The example of the regenerative crustacean challenged Descartes's stark division between body and mind, material and mental processes. While physical matter could be divided, cut, and quantified, the mind was indivisible, immaterial, and invisible. In addition, a rather arbitrary phenomenon like regeneration in crayfish called into question Descartes's denial of intelligent minds in animals. More specifically still, in the old Cartesian model of the animal machine, life depended on the organization of its parts. Such a system inevitably broke down when marshaled to explain the rebirth of such complex, living organs like crayfish limbs in ever more, in the words of Hartsoeker, "organized bodies."[108] The regeneration witnessed in crayfish centered on this fundamental difference in matter and mind, body and soul. Because crayfish regrew their limbs, Hartsoeker attributed the phenomenon to a conscious, intelligent mind rather than to a mute, inanimate bodily machine. Though not demonstrable, Hartsoeker's intelligent spirits that grew and learned various tasks over time in the bodies that they inhabited accommodated free will. At the same time, these Intelligences worked on God's behalf. Hartsoeker hoped he could thus resolve God's constant action in creation and propose a less deterministic and a more all-encompassing cosmology. What mattered to him was that leg regeneration in crayfish was a physical fact that offered support for a newer version of his physical system: that of an intelligent soul that organized matter in the body.

[108] Hartsoeker, "Remarques sur une These de Physique...," *Recueil de plusieurs pieces de physique*, 196–197. See the expression also in the title of his article in the *Bibliothèque ancienne et moderne* 18 (1722).

10

Epigenesis as Spinozism in Diderot's Biological Project

CHARLES T. WOLFE

1. Introduction

Denis Diderot's natural philosophy is deeply and centrally 'biologistic': As it emerges in the mid-18th century, thus right before the appearance of the term 'biology' as a way of designating a unified science of life, his project is motivated by the desire both to understand the laws governing organic beings and to emphasize, more 'philosophically,' the uniqueness of organic beings within the physical world as a whole. In what follows, I examine a little-known aspect of Diderot's articulation of his project: his statement in favour of biological epigenesis within the short but suggestive *Encyclopédie* article "Spinosiste."[1]

What possible relation could there be between Spinozism and epigenesis? Between a metaphysics of substance and modes that, even if it is also a major statement of philosophical naturalism, says almost nothing about biological entities, and a fashionable embryological theory of the 17th and 18th centuries?[2]

[1] Denis Diderot, "Spinosiste," *Encyclopédie*, vol. XV (1765): 474a. On the emergence of biology as such, see Joseph Caron, " 'Biology' in the Life Sciences: A Historiographical Contribution," *History of Science* n° 26 (1988): 223–268; Peter McLaughlin, "Naming biology," *Journal of the History of Biology*, vol. 35 (2002): 1–4; and Guido Barsanti, "Lamarck: Taxonomy and Theoretical Biology," *Asclepio*, vol. 52, n° 2 (2000): 119–131.

[2] Sometimes the *conatus* is presented as a 'vital force,' a 'survival principle' within the organism that leads it to seek to persevere in its existence; or Spinoza is interpreted as an 'organicist' (notably in Hans Jonas's "Spinoza and the Theory of Organism," *Journal of the History of Philosophy*, vol. 3 (1965): 43–57; reprinted in Jonas, *The Phenomenon of Life. Towards a Philosophical Biology* [New York: Harper & Row/Dell, 1966]). But, as François Duchesneau showed, it is a mistake to make such a sharp distinction between Cartesian mechanism and Spinozism: The conatus is itself a mechanical relation between the activity of one individual and others (Duchesneau, "Modèle cartésien et modèle spinoziste de l'être vivant," *Cahiers Spinoza* 2 [1978], 273).

A difficulty in many histories of biology, or works focusing on the so-called history and philosophy of biology (such as Grene and Depew's "episodic history"[3]) is that they flatten out the series of theories, positions, and controversies therein, in a rather linear fashion: preformationism versus epigenesis, Harvey versus Descartes on circulation, Bernard on the *milieu intérieur*, and so on. In contrast, my concern is not so much to stress innate complexity or the nonlinearity of any particular context (whether in the form 'Paracelsus was right!' or the counterfactual 'What if midwives had succeeded in their efforts to be recognized and form a Royal College?'), but to point out—here, with respect to epigenesis in its 'materialist' appropriation—that other factors, interests, and intentions are at work that do not fit well either into a history of discoveries or a catalogue of theoretical background positions implicit in the naturalist's practice. These sorts of factors are both ideological and metaphysical; they are often highlighted, in contrast, in histories organized around ideas such as 'the radical Enlightenment.'[4] But such histories have very little to say about the integral relation between such radicalism and shifts in the life sciences, for they focus on heterodoxy, politics, and of course philosophical polemics at the expense of naturalistic concerns (an exception being Ann Thomson's work, which seeks to tell a more unified story[5]). To be fair, the existence of a 'vital materialism' has been emphasized in recent scholarship (although the term goes at least as far back as the 1960s, with Jean Wahl and Yvon Belaval[6]). But this still leaves out the clandestine, radical dimension: The fascination with generation, species, or 'vital minima' is neither just the expression of prodromes or rough drafts of a future normalized science (as presented, typically, in works such as *Forerunners of Darwin*[7]), nor a merely ideological construct 'on top of' historical, socio-cultural discourses.

Epigenesis has been many things to many people. Most generically, it is the embryological theory that "organs . . . are progressively formed from, or emerge

[3] Marjorie Grene and David Depew, *The philosophy of biology: an episodic history* (Cambridge, UK: Cambridge University Press, 2004).

[4] Jonathan Israel, *Radical Enlightenment. Philosophy and the Making of Modernity, 1650–1750* (Oxford, UK: Oxford University Press, 2001).

[5] Ann Thomson, *Bodies of Thought: Science, Religion, and the Soul in the Early Enlightenment* (Oxford, UK: Oxford University Press, 2008).

[6] Hanns Peter Reill, *Vitalizing Nature in the Enlightenment* (Berkeley, CA: University of California Press, 2005). Jean Wahl describes Diderot as a "vitalist materialist" and the Leibnizian philosopher of nature Jean-Baptiste Robinet as a "materialist vitalist" (*Tableau de la philosophie française* [Paris: Gallimard, 1962], 53, 54). Yvon Belaval suggests that Diderot's non-mechanistic materialism, which brings him close to vitalist insights (of the non-supernaturalist variety), should be described as "un vitalo-chimisme ou un chimio-vitalisme"; "Sur le matérialisme de Diderot," in *Europäische Aufklärung. Herbert Dieckmann zum 60. Geburtstag*, eds. H. Friedrich and F. Schalk (Munich, DE: Wilhelm Fink, 1967), 9–21; reprinted in Belaval, *Études sur Diderot* (Paris: PUF, 2003), 367.

[7] *Forerunners of Darwin, 1745–1859*, eds. Bentley Glass, Owsei Temkin, and W. L. Straus (Baltimore, MD: Johns Hopkins University Press, 1959).

from, an originally undifferentiated, homogenous [material]."[8] Or, that organisms in development increase in complexity as the result of the operation of physical laws on a nexus of simpler resources.[9] But from Harvey to Maupertuis and Diderot (the story gets complicated by Kant, because he uses epigenesis both as a metaphor for his view of the development of the mind [i.e., that the origin of cognitions cannot be accounted for in either empiricist or innatist terms][10] and more literally, in a complex *Auseinandersetzung* with embryologists such as Johann Blumenbach, where Kant comes out on the side of the epigenetic theory in biology but worries about hylozoism[11]), epigenesis does count as a name for a point of intersection between a more *empirical* theory of biological development and a more *speculative* theory of the vital potentiality of matter to self-organize. In that sense, to focus on the case of epigenesis in just about any context in the 17th or 18th centuries is to be confronted with a total breakdown of any convenient distinction between 'experimental' and 'speculative' modes of natural philosophy (of the sort suggested, e.g., in Peter Anstey's recent work[12]). And this is not only true of materialists such as La Mettrie and Diderot; it is also the case in the reflections on generation and reproduction of Charles Bonnet.

[8] C. U. M. Smith, *The Problem of Life: An Essay in the Origins of Biological Thought* (New York: Wiley, 1976), 264.

[9] Paul Griffiths and Karola Stotz, "How the mind grows: a developmental perspective on the biology of cognition," *Synthese* 122 (2000), 34.

[10] "Intuitions of the senses (in accordance with sensible form or matter) yield synthetic propositions which are objective. Crusius explains the real principle of reason according to a systematae praeformis (from subjective principiis); Locke according to influxo physico like Aristotle; Plato and Malebranche from intuitu intellectuali; we according to epigenesis from the use of natural laws of reason" (a 'Reflexion' of 1770–1771, n° 4275 in Kant, *Kants gesammelte Schriften. Hrsg. von der Königlich Preussischen Akademie der Wissenschaften zu Berlin.* (Reprint, Berlin: De Gruyter, 1900–), XVII, 492. On epigenesis in Kant, see Wayne Waxman, *Kant's Model of the Mind* (Oxford, UK: Oxford University Press, 1991), 249–267.

[11] *Critique of Judgment* §81; *Kants gesammelte Schriften*, V, 424f.; on hylozoism: "We perhaps approach nearer to this inscrutable property if we describe it as an analogue of life, but then we must either endow matter, as mere matter, with a property that contradicts its very being (hylozoism) or associate it with a foreign principle standing in communion with it (a soul)" (ibid., §65; *Kants gesammelte Schriften*, V, 374–375). This can sound 'vitalistic,' which is a danger for a Kantian. Kant cites Maupertuis's molecules "endowed with intelligence" (and desire, aversion, and memory) as a major example of the dangers of hylozosim in *Der einzig mögliche Beweisgrund zu einer Demonstration des Daseins Gottes*, cited in John Zammito, "Kant's early views on epigenesis: The role of Maupertuis," in *The problem of animal generation in early modern philosophy*, ed. J. E. H. Smith (Cambridge, UK: Cambridge University Press, 2006), 343. On the metaphysics of Maupertuis's molecules, see C. T. Wolfe, "Endowed molecules and emergent organization: the Maupertuis-Diderot debate," *Early Science and Medicine* 15 (2010): 38–65.

[12] Peter Anstey, "Experimental Versus Speculative Natural Philosophy," in *The Science of Nature in The Seventeenth Century. Patterns of Change in Early Modern Natural Philosophy*, eds. P. Anstey and J. Schuster (Dordrecht, NL: Springer, 2005).

The question of genre (which of course is much more than a question of genre, but that is a convenient name for it), namely: When are we in the presence of a distinctively 'biological' idea, and when are we, in contrast, dealing with a more traditional and/or metaphysically founded 'matter theory'? is not an easy one, as for instance in the case of Francis Glisson and his metaphysics of life, a theory of innate potentialities or 'appetites' in matter.[13] Epigenesis is thus also a metaphysics of life before Diderot.[14]

In the case of William Harvey, who is considered to be the first to use the term 'epigenesis,' the blood exists first[15] and pulsates by a sort of fermentation, by 'an intimate heat or an innate spirit,' regulated by the *anima*; it is therefore the principal element in the body and the seat of the *anima*, and 'that in which heat, the primary and immediate instrument of life, is innate.'[16] Walter Pagel called this *vitalism*—rather anachronistic terminology in any case, whether we want to restrict ourselves to 'actors' categories' or, more common-sensically, wish to limit the usage of 'vitalism' to contexts in which a special vital property or arrangement is posited; Pagel speaks rather vaguely of "Aristotle's conception of the vital principle, the Anima,"[17] and of Harvey's vitalism; in both cases, there is no

[13] See Guido Giglioni, "Anatomist Atheist? The 'Hylozoistic' Foundations of Francis Glisson's Anatomical Research," in *Religio Medici: Medicine and Religion in Seventeenth-Century England*, eds. Ole Peter Grell and Andrew Cunningham, 115–135 (Aldershot, UK: Scolar Press, 1996); "What Ever Happened to Francis Glisson? Albrecht Haller and the Fate of Eighteenth-Century Irritability," *Science in Context*, vol. 21 (2008): 465–493; and Antonio Clericuzio, "The Internal Laboratory: the Chemical Reinterpretation of Medical Spirits in England (1650–1680)," in *Alchemy and Chemistry in the 16th and 17th Centuries*, eds. P. Rattansi and A. Clericuzio (Dordrecht, NL: Kluwer, 1994), 59.

[14] One can thus distinguish a Harveyan from a Glissonian 'tradition,' in which the latter articulates epigenesis with a metaphysics of life. This is for instance how the mortalist physician William Coward appropriates both Harvey and Glisson; Ann Thomson, "Encore l'âme matérielle," *La Lettre clandestine*, vol. 14 (2006), 64–65; and *Bodies of Thought: Science, Religion, and the Soul in the Early Enlightenment* (Oxford, UK: Oxford University Press, 2008), 113. Similarly, a metaphysics of vital properties—and the question of whether the physiologist should be concerned with it—appears in the controversy between Robert Whytt and Albrecht von Haller on irritability, in which Diderot appropriates both but sounds more like the former philosophically; François Duchesneau, "Diderot et la physiologie de la sensibilité," *Dix-huitième siècle* 31 (1999): 195–216; C. T. Wolfe, "Sensibility as vital force or as property of matter in mid-eighteenth-century debates," in *Sensibilité: The Knowing Body in the Enlightenment*, ed. Henry Martyn Lloyd (Dordrecht, NL: Springer, forthcoming).

[15] "Exercitationes de generatione animalium..." (1651), in *The Works of William Harvey*, trans. Robert Willis (London: Sydenham Society, 1847), 51, 52, 72; discussion in Alan Salter, "William Harvey. A Study in Empiricism," Ph.D. diss., University of Sydney, Unit for History and Philosophy of Science, 2010; and James G. Lennox, "William Harvey: Enigmatic Aristotelian of the 17th century," in *Teleology in the Ancient World: Philosophical and Medical Approaches*, ed. Julius Rocca (Cambridge, UK: Cambridge University Press, 2013).

[16] "Exercitationes de generatione animalium...," 52, in *Works*, 381, and cf. 373, 376, and all of 71, "On Innate Heat."

[17] Walter Pagel, "William Harvey: Some Neglected Aspects of Medical History," *Journal of the Warburg and Courtauld Institutes*, vol. 7 (1944), 147.

trace of a claim for the uniqueness of vitality as I previously defined it. Granted, Harvey does hold in the *Generation of Animals* that living things, "as soon as they are endowed with life... suffice for their own nourishment and increase, and this in virtue of peculiar inherent forces, innate, implanted from the beginning,"[18] and he speaks frequently in this work of a *vis plastica*.[19] He defines epigenesis as the "addition of the parts that successively arise," "the superaddition of parts... out of the power or potentiality of the preexistent matter (*ex potentia materiae pre-existentis*)."[20] But Harvey is not interested in extrapolating from his analysis of the formation of the egg to the metaphysics of life and matter itself, or better, even if he makes partly Aristotelian extensions from the embryological context toward the nature of life itself, his observations and interpretations are not founded on or interrelated with an ontology of life (in contrast to Glisson, but also to Diderot's 'Spinozist' construct that I discuss later).

2. Epigenesis as Materialism

How is it that this theory, epigenesis, becomes part and parcel of a vital materialism? For once we get to Maupertuis and Diderot, moving past the impasses reached by preformationist theory, whether ovist or animalist, we find older examples (e.g., ones given already by Harvey) being presented by Maupertuis in the *Vénus physique* in order to assert epigenesis, in the context of a conception of matter as endowed with vital, self-organizing properties: a vital, non-mechanistic materialism. We can call this position *materialism*, in that epigenesis as a theory recognizes in nature the power of self-formation as something material. But it was *vital* in that its proponents (notably Maupertuis) argued that the mechanical principles of physics did not suffice to account for embryo growth and the formation of organisms ("organized bodies") overall. Reflecting on this process of growth, Maupertuis stated that Newtonian attraction does not sufficiently account for organic phenomena, or even "the simplest chemical operations"[21]; this force alone cannot properly account for the production of *specifically organized bodies*: "A blind, uniform attraction distributed throughout the parts of

[18] Harvey, "De generatione animalium," Ex. 27, in *Works*, 281.

[19] Harvey, "De generatione animalium," preface: "De methodo in cognitione generationis adhibenda"; *vis plastica* is translated as "plastic force" or "plastic power" by R. Willis, in his 1843 edition of Harvey's works. Antonine Nicoglou has counted 23 references to '*vis plastica*' in this work. See her thesis in progress, "La plasticité du vivant: une analyse épistémologique," Université de Paris-1/IHPST.

[20] Harvey, "De generatione animalium," ex. 51, 45, in *Works*, 372, 335.

[21] Pierre-Louis Moreau de Maupertuis, "Système de la nature. Essai sur la formation des corps organisés and Réponse aux objections de M. Diderot," in *Œuvres*, 2 vols. (Lyon, FR: Bruyset, 1756), §III, 141.

matter would not explain how these parts arrange themselves to form even the simplest organized body.... Why shouldn't they unite at random? (*Pourquoi ne s'unissent-elles pas pêle-mêle?*)"[22]

Notice that if the theory of epigenesis now explicitly involves the capacity of matter to self-organize, so that complexity emerges out of material processes alone, we are no longer in a strictly biological context. In that sense, again reflecting elements of the story that we will not find in a history of biology (or biological thought), it is important to notice how scandalous the doctrine can be. Diderot's statement of epigenesis as Spinozism is my key case of this scandalousness, but examples can be found quite earlier. For instance, in the early 1700s, Samuel Clarke attacks Anthony Collins for this view, declaring that

> It being as impossible that the organized Body of a Chicken should by the power of any Mechanical Motions be formed out of the unorganized Matter of an Egg; as that the Sun Moon and Stars, should by mere Mechanism arise out of a Chaos.[23]

Collins had argued that

> the Matter of which an Egg consists, doth intirely constitute the young one, and that the Action of Sensation began under a particular Disposition of the Parts by Motion, without the Addition of an Immaterial and Immortal Soul, as the Powers of Vegetation, Gravitation, of producing the Sensation of Heat, Cold, Red, Blue, Yellow, are performed without the Addition of an Immaterial and Immortal Soul.[24]

As with Harvey and Glisson, it is not easy here to separate the strictly medical or 'biological' content of a theory and the ideologically charged polemical (but also metaphysically speculative) dimension. This is also apparent in Thomas Willis's 'pyrotechnic' account of generation. Willis was only trying to bring together chemistry, anatomy, and physiology to produce an integrated model of brain function and cognitive processes, without materialist intentions, but Henry More attacked him for what he called 'Psychopyrism'[25]: He felt that "according to

[22] Maupertuis, "Système," §XIV, 146–147; and Wolfe, "Endowed molecules and emergent organization."

[23] Clarke, "Letter to Mr. Dodwell [for Anthony Collins]" (Second Defence of the Immateriality and Immortality of the Soul), in *The Works of Samuel Clarke*, 4 vols., 1738 (reprint, New York: Garland, 1978), III, 789.

[24] Collins, "A Reply to Mr Clarke's Defence of his Letter to Mr Dodwell," in Clarke, *Works*, III, 768.

[25] See his 1682 "A Letter to a Learned Psychopyrist"; John Henry, "Medicine and Pneumatology: Henry More, Richard Baxter, and Francis Glisson's Treatise on the Energetic Nature of Substance," *Medical History*, vol. 31 (1987), 34.

[Willis] the production of a Soul ex Traduce, would end in meer Materialism,"[26] with the explicitly epigenetic dimension being that "in Generation some matter only is newly modified."[27] And indeed, Willis had spoken of the "Vital Humour in an Egg."[28] In the 18th century, we also find epigenetic concepts either being used with deliberate *philosophical*-materialist overtones, or as experimental/ medical concepts, again with some overlap between the two. The physician Abraham Gaultier's semi-clandestine treatise *Parité de la vie et de la mort* (1714) puts forth an emergentist concept of matter, medically inspired and with frequent recourse to epigenetic concepts.[29] When the vitalist physician Théophile de Bordeu reflects on the nature of 'cellular substance,' he describes a gradual shift from a kind of vital glue, a mucous, nutritive substance, to sheaths of muscular fibre, solely by material superaddition.[30]

To take stock for a moment, these various examples, from More on Willis and Clarke on Collins to later medical theories and the debate between Maupertuis and Diderot, show (i) that a theory about generation is intertwined with shifts in matter theory, and (ii) that this intertwinement has a radical dimension, sensu Israel (whose analysis of the Radical Enlightenment,[31] however, tends to steer clear of the sciences)—for epigenesis is not a metaphysically dangerous theory when Harvey introduces it in 1651. There are other, better-known cases that also combine these elements: The example of the polyp, which fascinated a generation of European scientists, was explicitly taken by Charles Bonnet as an exciting but dangerous challenge to the existence of the soul, or at least of a single, indivisible soul in the body. One could also mention the anatomico-metaphysical study of monsters, or the concern with species and miscegenation in this period.[32] But the articulation of epigenesis and materialism shows us something else again at this intersection of radicalism and the new focus on the emergent life sciences. And the sharpest, most distinctive form this takes is in Diderot's Spinozism.

[26] Joseph Glanvil and Henry More, *Sadducismus triumphatus: or, A full and plain evidence concerning witches and apparitions* (London: A. Bettesworth and J. Batley, 1726 [1681]), 129.

[27] Ibid., 130.

[28] Thomas Willis, "Two Discourses Concerning the Soul of Brutes, Which is That of the Vital and Sensitive [Soul] of Man. The first is physiological, shewing the nature, parts, powers, and affections of the same. The other is pathological, which unfolds the diseases which affect it, a translation of De anima brutorum" (1672), English trans. S. Pordage (London: Dring, Harper and Leigh, 1683), 33.

[29] Abraham Gaultier, *Parité de la vie et de la mort. La Réponse du médecin Gaultier (1714)*, ed. O. Bloch (Paris: Universitas; Oxford, UK: Voltaire Foundation, 1993).

[30] Théophile Bordeu, "Recherches sur le tissu muqueux" (1767), §VII, in Bordeu, *Œuvres complètes*, 2 vols. (Paris: Caille et Ravier, 1818), II, 736.

[31] Israel, *Radical Enlightenment*.

[32] Charles T. Wolfe, ed., *Monsters and Philosophy* (London: Kings College Publications, 2005); and Justin E. H. Smith, ed., *The Problem of Animal Generation in Early Modern Philosophy* (Cambridge, UK: Cambridge University Press, 2006).

3. Diderot's Biological Project

Diderot's biological project is inseparable from claims that are not themselves restrictively biological; he is the proponent of a (programmatic) Spinozist biology. Diderot was not a physician like La Mettrie, or a 'working natural historian' like Buffon; but one of his first publications was the translation of Robert James's *Medicinal Dictionary* (1745), and in addition to his enormous activity as the chief editor of the *Encyclopédie* (which heavily features medical entries, sometimes with his editorial interventions), he was also a serious student of chemistry, including 'vital chemistry.'[33] And in the *Éléments de physiologie*, the manuscript on which he worked during the last two decades of his life, he asserts: "Pas de livres que je lise plus volontiers que les livres de médecine."[34] In fact, as its title indicates, this work is about the elements of *physiology* (understood in the broad sense as a study of living animals and humans, as notably opposed to anatomy). Rather than having a specifically medical or medico-philosophical focus, Diderot is concerned with 'natural history,' by which he means the study of the nature of life as a whole, including its ontological status (whereby natural history has a specifically materialist dimension[35]).

The latter dimension is apparent when Diderot moves within one sentence, as in his speculative, experimental work the *Rêve de D'Alembert* (1769), from a statement of epigenesis in the restrictive sense, to claims such as "Do you see this egg? It is with this egg that we can overturn all schools of theology."[36] This idiosyncratic combination comes in different prose forms in Diderot: sometimes in enlightened commentaries on experimental science (*Pensées sur l'interprétation de la nature, Principes philosophiques sur la matière et le mouvement, Éléments de physiologie*), sometimes in more speculative forms, which are harder to categorize (particularly the *Lettre sur les aveugles* and *Le Rêve de D'Alembert*); the latter has been described as Diderot's 'science-fiction.' In both of these sorts of works, Diderot is haunted or rather fascinated by the nature of living beings—their capacity to transform, to produce monsters, to return to life when the substance appeared dead, in short, their "polypous"[37]

[33] Diderot attended Guillaume-François Rouelle's chemistry lectures at the Jardin du Roi (now the Jardin des Plantes) between 1754 and 1757, and served as secretary during many of the lectures, taking notes and even preparing some of the lectures for Rouelle, which were first published as "Introduction à la chymie, manuscrit inédit de Diderot publié avec une notice sur les cours de Rouelle," ed. Charles Henry (Paris, 1887); now available as "Cours de chimie de Mr Rouelle," in Diderot, *Œuvres complètes*, eds. H. Dieckmann, J. Proust, and J. Varloot (Paris: Hermann, 1975–), vol. IX.

[34] Diderot, *Éléments de physiologie*, in *Œuvres complètes*, vol. XVII, 510.

[35] Charles T. Wolfe, "'Cabinet d'Histoire Naturelle,' or: The Interplay of Nature and Artifice in Diderot's Naturalism," *Perspectives on Science*, vol. 17, n° 1 (2009): 58–77.

[36] Diderot, *Œuvres complètes*, XVII, 103–104.

[37] See Jacques Proust, "Diderot et la philosophie du polype," *Revue des sciences humaines*, vol. 14, n° 182 (1981).

nature, in all its "vicissitudes" (a term Diderot uses in a deliberately Lucretian manner, throughout his work, to mean a kind of perpetual flux and transformation).

So on the one hand, Diderot's interactions with the life science of his time can be understood in a straightforward sense as the activity of an educated individual with a strong interest in the implications for philosophy and traditional knowledge overall of new discoveries or conceptual schema, whether from medicine (with implications for knowledge about behavior), biology (implications for questions of reproduction and identity), or natural history (implications for the status of species and evolution). But on the other hand, his articulation of all of these in a *materialist project* does not belong in or open onto an episode among others in the history of science. I shall try to make both of these aspects more clear by (i) situating Diderot's concern with 'biological life' in the context of the emergence of biology as a science, and then (ii) turning to what I term here his 'Spinozist biology.'

3.1. Diderot's Biologism and the Emergence of 'Biology'

In a dramatic section of his *Pensées sur l'interprétation de la nature* (1753–54), Diderot seems to announce, more directly than in his usual, more metaphorical style, the advent of something like *biology*:

> We are on the verge of a great revolution in the sciences. Given the taste people seem to have for morals, *belles-lettres*, the history of nature and experimental physics, I dare say that before a hundred years, there will not be more than three great geometricians remaining in Europe. The science will stop short where the Bernoullis, the Eulers, the Maupertuis, the Clairaut, the Fontaines and the D'Alemberts will have left.... We will not go beyond.[38]

This passage has rarely been commented on, and scholars who have done so tend to miss the radical dimension. Thus Paolo Casini only notices that Diderot is mistaken in his diagnosis of the situation of mathematics: "This was a rather curious reaction in a period when the best continental mathematicians were active in solving so many problems left open in the *Principia*." Somewhat closer to the point (but only in part), Casini does add that "at this time the life sciences entered a period of rapid growth and needed a proper new method."[39] Indeed, similar proclamations about the rise of life science *together* with a denial

[38] *Pensées sur l'interprétation de la nature*, §IV, in Diderot, *Œuvres complètes*, IX, 30–31.

[39] Paolo Casini, "Newton's 'Principia' and the Philosophers of the Enlightenment," *Notes and Records of the Royal Society of London* 42:1 (1988), special issue on Newton's 'Principia' and Its Legacy, 44.

of the pertinence of the mechanical, physical, and mathematical sciences can be found elsewhere, including the central article of the *Encyclopédie*, entitled "Encyclopédie": "There is a general trend in people's minds towards natural history, anatomy, chemistry and experimental physics."[40] Similarly, Buffon stated in the first volume of his *Histoire naturelle* that "mathematical truths are just abstractions of the mind, that are in no way real."[41]

This *is* science, or programmatic science, although it is anti-mathematical (and to be clear, Buffon the translator of Stephen Hales's *Vegetable Staticks* and Diderot the author of several essays on probabilities, were by no means mathematically ignorant). Another example would be the medical vitalist focus on properties of organs (or muscles, such as irritability) that cannot be grasped mathematically (or at least are the object of failed calculations by iatromathematical physicians such as Keill and Borelli); a medical thesis on irritability defended at Montpellier in 1776 by a certain Mr. 'D.G.' (who further research identifies as Jean Charles Marguerite Guillaume de Grimaud), discusses properties of muscles but also 'animals' such as the polyp—both in terms of their contractility—in order to prove that the "human machine," the "most complicated machine in the universe," is not reducible to the simple, atomic, intellectual objects of mathematics.[42] But Diderot is neither a natural historian nor a professor of medicine. What is his relation then to these proclamations of the supremacy of the life sciences? Is he putting forth essayistic sketches of a proto-biology, or a philosophical materialism with a vital flavour, as indicated before? Let us consider each possibility in turn.

If we read Diderot as somehow an actor in the emergence of biology, 40-odd years before its official appearance (or an observer of this appearance, whether or not he was right in his predictions about geometry), the reader may object that this is anachronistic. Yet I do not think this is the real problem with such an identification. The standard view is that 'biology' as a term appears in the late 1790s,

[40] *Encyclopédie*, in Diderot, *Œuvres complètes*, VII, 185. Analogous passages can be found in Diderot's letter to Voltaire of Feb. 19, 1758, and in Grimm's *Correspondance littéraire*, June 1, 1765, vol. IV, 1, 649. The *Encyclopédie* article "Histoire naturelle," which is primarily by Diderot with some excerpts of Buffon, contains a reflection on trends and "fashions in the sciences"; Diderot notes that the "taste for abstract and mechanical sciences" replaced the taste for the study of antiquity, and that the former taste was in turn replaced by the taste for "experimental science," which itself is now losing ground to "natural history" (an umbrella term for life science partly interchangeable with 'biology,' as I discuss at greater length in "Cabinet d'Histoire Naturelle"). But given this *hauteur de vue*, Diderot acknowledges—unlike thinkers we might think of as 'vitalists' or 'organicists'—that there is no reason why this biocentric focus might itself some day not be replaced by something else ("le règne de l'histoire naturelle aura-t-il aussi son terme ?") (*Encyclopédie* VIII, 1765, 228b).

[41] Georges-Louis-Marie-Leclerc de Buffon, "De la manière d'étudier l'Histoire Naturelle," in *Histoire naturelle, générale et particulière*, vol. 1 (Paris: Imprimerie Royale, 1749), 53.

[42] 'D.G.' (Jean-Charles-Marguerite-Guillaume de Grimaud), *Essai sur l'irritabilité* (Avignon, FR: Bonnet frères, 1776), 33, 35.

in works by Treviranus and Lamarck (roughly at the same time)[43]; Lamarck planned for a long time to write a treatise entitled *Biologie, ou Considérations sur la nature, les facultés, les développements et l'origine des corps vivants* (in fact, some of the manuscript, dated 1800, survives, but it did not circulate; it was published for the first time in 1944).[44] Treviranus aimed not only to provide a precise compilation of the knowledge of his time, concerning the phenomena of life, but also a theoretical framework for this new scientific discipline he called biology, by combining philosophical and experimental analysis and information.[45] Less well known is that the term 'biology' occurs in the context of *Naturphilosophie*, in authors such as T. G. A. Roose, K. F. Burdach, and C. C. E. Schmid,[46] or that there is a good deal of terminological instability in the decades prior to 1800, whether it is the presence of *biologi* in Latin, but with a different meaning (scientists studying plants[47]), or the presence of competitor terms such as 'zoonomia' or 'biogeography' well until the 1830s.

But all of this confusion or profusion—semantic, scientific, conceptual—does, however, reveal an increasing concern with an ontological domain (including in all its colorful, popular images: spontaneous generation, polyps, monsters, and so on); and Diderot is certainly part of this concern, and indeed an important actor in its dissemination,[48] including the transformations of what had been 'natural philosophy' and the role of the philosopher therein. However, this does not capture his real motivations: He is not seeking to be a more materialist version of a Treviranus or a *Naturphilosoph*. If we understood him thus,

[43] G. R. Treviranus, *Biologie oder Philosophie der lebenden Natur für Naturforscher und Aerzte*, 6 vols. (Göttingen, DE: J. F. Röwer, 1802–22); and J.-B. de Lamarck, *Hydrogéologie, ou Recherches sur l'influence qu'ont les eaux sur la surface du globe terrestre* (Paris: Agasse & Maillard, 1801–02).

[44] Jean-Baptiste de Lamarck, "Biologie ou Considérations sur la nature, les facultés, les développemens et l'origine des corps vivans," ed. P.-P. Grassé , *La Revue Scientifique* 82 (1944), 267–276.

[45] Elke Witt, "Die wechselnden Gewänder der Natur: Die Biologie nach Gottfried Reinhold Treviranus," in M. Kaasch, J. Kaasch, and N.A. Rupke, eds., *Physische Anthropologie—Biologie des Menschen* (Berlin: VWB-Verlag 2007), 177–186.

[46] T. G. A. Roose, *Grundzüge der Lehre von der Lebenskraft* (Braunschweig, DE: Christian Friedrich Thomas, 1797); C. C. E. Schmid, *Physiologie philosophisch bearbeitet*, 3 vols. (Jena, DE: Akademische Buchhandlung, 1798–1801); and K. F. Burdach, *Propädeutik zum Studium der gesammten Heilkunst* (Leipzig, DE: n.p., 1800). On the development of biology, including as a self-conscious discipline in these years (1795–1802), see the sources cited earlier: Barsanti; Caron; and McLaughlin. For the issue of 'before and after' the naming of 'biology,' see Charles Wolfe, "Why was there no controversy over Life in the Scientific Revolution?" in V. Boantza and M. Dascal, eds., *Controversies in the Scientific Revolution* (Amsterdam: John Benjamins, 2011), 187–219.

[47] Robert Richards, *The Romantic Conception of Life: Science and Philosophy in the Age of Goethe* (Chicago: University of Chicago Press, 2002), 4n.

[48] Thus the *Encyclopédie* devotes much more room to the life sciences than any predecessor work does, as discussed by Claire Salomon-Bayet, *L'institution de la science et l'expérience du vivant: méthode et expérience à l'Académie Royale des Sciences, 1666–1793* (Paris: Flammarion, 1978).

we would be neglecting the 'Spinozist' dimension of his thought, which is both more speculative and more radical. I've already noted that Diderot's interest (or even fascination) in the nature of living beings, *does not open onto biology as a (nascent) science*, and indeed, does not aim at this development. As can be seen in the article "Spinosiste" and the other texts cited later, Diderot is articulating a junction, a connection among traditional metaphysical considerations, experimental revelations on the nature of life, and a new kind of philosophical project that is neither presenting itself as the *foundation* of the study of the natural world nor as a friendly *ancillary* to such study.

3.2. Diderot's Spinozist Biology

So what is Diderot's Spinozist biology, or biologistic Spinozism? It is positively articulated in two texts of a very different nature: his short article "Spinosiste" in the *Encyclopédie* (which is the most 'academic' presentation of the present theme, but also perhaps the most surprising) and a more speculative version of the same text, which we find in his 'experimental' work *Le Rêve de D'Alembert*; it is negatively or 'discursively' articulated in his polemic with Maupertuis, which occurs some 10–15 years earlier, but which I discuss last.[49]

Here is the definition of a Spinozist given in the *Encyclopédie* (not to be confused with the long, more conventional entry on Spinoza, which is largely authored by the Abbé Yvon):

> SPINOSIST, s. m. (*Gram.*): follower of the philosophy of Spinosa. One must not confuse the ancient *Spinosists* with the modern *Spinosists.* The general principle of the latter is that matter is sensitive; they demonstrate this by the development of the egg, an inert body which by the sole means [*instrument*] of graduated heat moves to the state of a sensing, living being, and by the growth of any animal which in its inception [*principe*] is merely a point, and through the nutritive assimilation of plants and—in one word—of all substances that serve the purpose of nutrition, becomes a great sensing and living body in a greater [expanse of] space. From this they conclude that only matter exists, and that it is sufficient to explain everything. For the rest, they follow ancient Spinosism in all of its consequences.[50]

The juxtaposition (or articulation) is surprising: There are ancient and modern Spinozists, and what characterizes the latter is essentially their commitment to epigenesis. Now, Diderot was fascinated by "the development of the egg" and related biological phenomena throughout his work, and he relates these to more

[49] Wolfe, "Endowed molecules."

[50] Diderot, "Spinosiste," *Encyclopédie*, vol. XV, 474a.

metaphysical questions: the nature of matter, the possibility that all matter can sense ("is sensitive"), and the shift from a state of inertia to a state of sensibility, or from inert matter to sensing matter, as we see in the above passage and also in those from the *Rêve* that I quote below. I hope it is clear that Diderot sees this as an inseparable relation between empirical phenomena (e.g., "here are two quite common phenomena: . . . the development of the egg, this move from the state of inert matter to the state of sensing being . . . the return to life of some insects"[51]) and a materialist doctrine in which the innate 'vitality' of matter demonstrates, or should demonstrate, the vacuity of any purported distinction between matter and thought, or matter and life. This *vital* dimension of Diderot's materialism is well known, along with the role played in his thought by the shifts—epistemological, experimental, and other—in the 'proto-biology' of his time.

But why present this claim in the article "Spinosiste"? In fact, very few commentators have asked *why* Diderot gives such an idiosyncratic definition of "modern Spinozism." To be sure, his convictions regarding living matter (or all of matter inasmuch as it is potentially living and sensing) are tied to his admiration for the metaphysics of a single substance composed of an infinite number of modes ("There is only one substance in the universe," he states in the *Rêve*[52]). But nowhere does Spinoza seek to connect his metaphysics to the life sciences; even if the notion of the conatus was frequently taken up in the generations after him to mean something like a survival impulse in living beings, this was not what he meant at all.[53] One of the few writers who did address my question (Why should epigenesis be presented as the view of modern *Spinozists*?), Paul Vernière, invented a category meant to cover such cases: "neo-Spinozism" (which was partly controversial). He defined this as a form of holist materialism *founded on the life sciences* rather than on a priori metaphysical speculation: "The neo-Spinozists are not abstract speculators but *savants*; starting from precise experiments on embryogenesis and animal physiology, they claim to find in matter itself the laws governing the origin and development of life."[54] The mystery surrounding the article

[51] Diderot, "Observations sur Hemsterhuis," in *Œuvres*, vol. 1: Philosophie, ed. L. Versini (Paris: Laffont-Bouquins, 1994), 708.

[52] *Rêve*, in *Œuvres complètes*, XVII, 107.

[53] For an interesting way of reading Spinoza as a thinker of 'Life,' however, see Sylvain Zac, *L'idée de vie dans la philosophie de Spinoza* (Paris: PUF, 1963), esp. ch. IV (as indicated earlier, Hans Jonas's essay "Spinoza and the Theory of Organism" is less useful here). For the repercussions of 'Spinozism' as heuristic but also polemical construct in Enlightenment medicine and biophilosophy (e.g., in Boerhaave), see Annie Ibrahim, "Sur le spinozisme dans les philosophies du vivant," in *Spinoza au XVIIIe siècle*, ed. O. Bloch (Paris: Klincksieck, 1990), 121-132; and for a more measured historical assessment, see Giglioni, "Whatever happened to Francis Glisson?" 485–486.

[54] Vernière, *Spinoza et la pensée française avant la Révolution*, 2nd edition (Paris: PUF, [1954], 1982), 529; he also describes Diderot (and Maupertuis) as neo-Spinozists in an editorial note to his edition of Diderot, *Œuvres philosophiques* (Paris: Garnier, 1961), 229. For more extensive

"Spinosiste" would then be dispelled: to be a neo-Spinozist is to seek to rearticulate a new form of (metaphysical) monism that would be in closer agreement with scientific data. But there is something odd about this definition, perhaps oddly neat in its way of demarcating a new form of science-friendly philosophical practice. For Diderot is quite willing to take statements of epigenesis into new territories that are neither inductively nor otherwise experimentally founded.

We can see this most clearly in one of his masterpieces, the experimental philosophical dialogue entitled *Le Rêve de D'Alembert* (*D'Alembert's Dream*) written in 1769, one copy of which was given by Diderot to Catherine the Great as a present. It is composed of three dialogues, of which the first two concern us here: In the first, the character Diderot debates the character D'Alembert on the nature of matter, thought, and sensibility, and tries to convince the latter that all of these are really on a continuum. Shortly after D'Alembert has said to Diderot "you have something against the distinction between the two substances," Diderot asks, rather rhetorically,

> Do you see this egg? It is with this [egg] that we overturn all schools of theology and all the temples of the world. What is this egg? An unsensing mass prior to the introduction of the seed [*germe*]; and after the seed has been introduced, what is it then? An unsensing mass, for the seed itself is merely an inert, crude fluid. How will this mass move to another [level of] organization, to sensibility and life? By means of heat. What will heat produce therein? Movement.[55]

A lot is happening in this passage. To claim that "it is with this [egg]" that "all schools of theology" can be overturned is obviously to step outside of the controlled, empirical claims of biological science. We could call this an ideological or polemical moment (precisely illustrating what we have come to call the Radical Enlightenment); it is also, of course, a step toward philosophical materialism. Furthermore, it is a step away from 'factual' claims about one system of generation (epigenesis) versus another (preformationism) toward a metaphysics of living matter: In this integrated, causally closed universe, which is composed of "only one substance,"[56] there is no particular demarcation between dead and living matter. And in case we still thought we were dealing with neo-Spinozism understood as a kind of scientifically grounded materialist metaphysics, Diderot happily states here and elsewhere that we are dealing with

discussions of Diderot's usage of Spinoza and Spinozism, see Alexandre Métraux, "Über Denis Diderots physiologisch interpretierten Spinoza," *Studia Spinozana* n° 10 (1994), 121-134; and John Zammito, "Naturalizm XVIII Wieku. Spinozyzm w Filozofiach nauki Diderota i Herdera," in *Rozum i świat: Herder i filozofia XVIII, XIX i XX wieku*, eds. Marion Heinz, Maciej Potepa, and Zbigniew Zwolin'ski (Warsaw: Genessis, 2004), 117–146.

[55] Diderot, *Œuvres complètes*, XVII, 103–104.

[56] *Rêve*, in Diderot, *Œuvres complètes*, XVII, 108.

conjectures, suppositions, thought experiments, and all sorts of imaginative, speculative constructs. Crucially—for this sets him apart from proponents of a 'metaphysics of life' like Glisson or, differently, Stahl[57]—Diderot acknowledges that "the necessary connection in this shift [namely, from brute matter to thinking matter, or from matter to sensibility and thought, CW] escapes me."[58] As has frequently been noted, Diderot chooses to put forth some of his most original claims regarding matter, life, and sensibility in an experimental work that is neither a philosophical treatise nor an experimental scientific report.[59]

In the second dialogue of the *Rêve*, the character Bordeu twice tells Mlle de Lespinasse, when she is puzzled by some of the speculation (or by biological discussion), to "do in thought (*par la pensée*, literally 'by thought') what Nature does sometimes" (XVII, 149). Here is a case in which she is reporting one of the hallucinatory 'rêveries' or divagations of the dreaming or somnolent D'Alembert, where Diderot is essentially reconstructing and expanding the theory of epigenesis so that it becomes a theory of living matter in general:

> It is certain that contact between two living molecules is something different from the contiguity of two inert masses.... A thread made of pure gold...—a homogeneous network. Between its molecules, others interpose themselves and perhaps form another homogeneous network, a tissue of sensitive matter, a contact which absorbs active sensibility from here and latent sensibility from there and which communicates itself like motion, without including... that there must be some difference between the contact of two sensitive molecules and the contact of two molecules which are not, and this difference—what could it be?... a habitual action and reaction... and this action and this reaction with a unique character.... Everything concurs thus to produce a sort of unity which only exists in the animal...."[60]

[57] On Glisson, see Giglioni, "What Ever Happened to Francis Glisson?"; on Stahl as defending a form of vitalism, see Kevin (Ku-Ming) Chang, "From Vitalistic Cosmos to Materialistic World," in Lawrence M. Principe, ed., *Chymists and Chymistry. Studies in the History of Alchemy and Early Modern Chemistry* (Sagamore Beach: Watson Publishing International LLC, 2007), 215-225; on interrelations between and shifts from 'metaphysics of life' to 'biology,' see Charles Wolfe, "Why was there no controversy over Life in the Scientific Revolution?"

[58] *Réfutation d'Helvétius*, in Diderot, *Œuvres*, vol. 1, 798.

[59] Compare the attention to genre in Jean-Claude Bourdin, *"Du Rêve de D'Alembert aux Éléments de physiologie.* Discours scientifique et discours spéculatif dans *Le Rêve de D'Alembert," Recherches sur Diderot et sur l'Encyclopédie* 34 (2003): 45–69 with the more strictly 'scientific' focus in Duchesneau, "Diderot et la physiologie de la sensibilité," discussion in Wolfe, "Sensibility as vital force or as property of matter."

[60] Diderot, *Rêve de D'Alembert*, in *Œuvres complètes*, XVII, 119; translation (modified) from http://records.viu.ca/~johnstoi/diderot/dalembertsdream.htm.

It is not just that Diderot is describing the organizational, transformative potential of "contact" (which he elsewhere calls "continuity") between two living molecules in metaphorical terms ("a thread made of pure gold"). More originally, he is engaging in a form of scientific speculation, as we can see with his appeals to chemical concepts ("action and reaction"), in order to do justice to an apparently holistic phenomenon (the "sort of unity which only exists in the animal").

Diderot articulates a connection between Spinozism and epigenesis in two very different kinds of texts: an apparently academic, precise entry in the *Encyclopédie* that combines (or juxtaposes) these different dimensions in a surprising way, and a more speculative, experimental series of reflections in the *Rêve de D'Alembert.* But as I indicated, there is a third locus for this topic: his debate with Maupertuis a decade earlier regarding the basic units of living matter or vital *minima* ("molecules") and what metaphysics is, or should be implicit therein. Some ramifications of Spinozism as an item of or within natural philosophy (fleshing out the persona we encountered before of the 'modern Spinosist' described in the *Encyclopédie*) can be found in this debate, which is both about the units of life and how these should be understood and justified metaphysically: a Leibnizian pan-psychist vision of molecules possessing intentional properties (Maupertuis), versus an emergentist, organizational vision in which these properties are only the properties of the *Whole,* not of the elements. Put differently, the Maupertuis-Diderot debate concerning the nature of what they termed 'molecule' is in fact a debate over attribution of properties: Should these be applied to the element or the organizational whole? On the surface, this exchange or polemic is also a case of the two authors trading accusations of atheism and Spinozism with each other (for the definition of matter as possessing dynamic, organizational, indeed 'intellective' properties is of course a dangerous one). I briefly reconstruct the debate as regards the present topic.

In 1751, Maupertuis had published (supposedly in Erlangen, but actually in Berlin) a Latin treatise entitled *Dissertatio inauguralis metaphysica de universali naturae systemate,* under the pseudonym Dr, Baumann, which he translated into French in 1754 and published with a more specifically 'biological' title: *Essai sur la formation des corps organisés.*[61] Diderot critically discussed the 'Erlangen dissertation' (and outed its author) in the second edition of his *Pensées sur l'interprétation de la nature* in 1754 (the first edition had appeared one year earlier), in sections L–LI. Maupertuis replied to Diderot's criticisms in a *Réponse*

[61] The final version of the text appeared in French with a title closer to the original, *Système de la nature,* in the 1756 edition of Maupertuis's *Œuvres.* Marx Wartofsky's 1952 paper is still one of the best commentaries on the topic ("Diderot and the Development of Materialist Monism," *Diderot Studies* n° 2 (1952): 279–329, later reprinted in his *Models. Representation and the Scientific Understanding, Boston Studies in Philosophy of Science,* vol. 48 (Dordrecht, NL: Reidel, 1979), specifically 292–293).

aux objections de M. Diderot included in the third and final version of his essay in his 1756 *Œuvres*.

In section XIV of the *Essai*, Maupertuis asked his readers to imagine a molecule "endowed with desire, aversion and memory."[62] Diderot rejected the idea of endowed molecules, at least in the essential form as stated by Maupertuis. In order to challenge Maupertuis's hypothesis, which he accepts at the level of its "empirical" benefits but not as an overall "speculative" claim, he seeks to push it as far as it can go, to its "terrible consequences."[63] (Maupertuis responded that if one were not already convinced of the sincerity of Diderot's religious beliefs, one might suspect that his intention is not so much to overturn his own theory as to "himself draw these consequences he calls *terrible* from it."[64]) Diderot felt that Maupertuis's molecules seemed to have been spiritualized, whereas he, Diderot, wanted to materialize the realm of the spiritual (i.e., the mental): First, and epigenetically, there is matter and motion, and gradually, through corpuscular arrangements of increasing complexity—which he terms "organization"—the phenomena or rather faculties of desire, aversion, memory, etc., are added on. Now, this seems rather mechanistic in contrast to Maupertuis; and indeed at this point, Diderot introduced the "dull sensitivity" (*sensibilité sourde*) by means of which all molecules have their place or fit into place (recall the "latent sensibility" of the molecules in the *Rêve de D'Alembert*). To be precise, Diderot allows for two properties: this rudimentary form of sensitivity, and an "automatic restlessness" (*inquiétude automate*) that leads the molecules into a variety of possible locations.[65]

The key component in Maupertuis's response to Diderot is a notion that is equally important in Diderot's own thought, that of the Whole (*le Tout*), which has explicit Spinozist overtones. In that sense, when the two authors accuse each other of being Spinozists, it is partly out of bad faith and polemical motivations, but also partly because two monistic visions of matter are at stake—and Diderot is trying to bring out the implicit radicalism of this vision, which Maupertuis wants to keep hidden, wearing the mask of the scientist. It is because both Maupertuis and Diderot are committed to a notion of the universe

[62] "Système," §XIV; the term 'endowed' appears in §§XXXI and LXVI. At the end of the book, Maupertuis speaks of "originarily endowed elements" (§LIV, 173). The "Système" and the "Réponse" are in vol. 2 of the *Œuvres*; cited as "Système" followed by section number (in Roman numerals), and "Réponse," followed by page number.

[63] From the outset (*Pensées sur l'interprétation de la nature*, §L, in Diderot, *Œuvres complètes*, IX, 77), Diderot had declared that he intended to push Maupertuis's hypotheses to their ultimate... or absurd conclusion, in order, he claims somewhat deceptively, to unmask the "terrible consequences" of the Erlangen doctor's theory.

[64] Maupertuis, "Réponse," 197.

[65] Diderot, *Pensées*, §LI, *Œuvres complètes*, IX, 84.

as a substantial Whole that they are 'neo-Spinozists' in a broad sense, to use Vernière's term—in addition to the specific, idiosyncratic sense manifest (e.g., in the article "Spinosiste") as a proponent of epigenesis who is also committed to substance metaphysics.

Yet as I indicated before, we have to be cautious in endorsing Vernière's concept with respect to both Maupertuis *and* Diderot, because he insists that what differentiates *neo*-Spinozists from Spinozists *tout court* is that the former base their reflections on scientific experiment and induction. This certainly does not apply to the—productive?—speculations in the *Rêve de D'Alembert*. Neither in that text, nor in the *Lettre sur les aveugles*, nor even in the *Éléments de physiologie*, does Diderot reason in strictly a posteriori 'empirical' terms, moving from facts to inductive generalizations. (One need only compare Diderot to Haller, whose physiological works he read very closely: What distinguishes them is precisely the Spinozist/radical element.[66]) In addition, the observations, practices, and experiments on which the epigenetic theory rests, in Diderot's context, are, as Olivier Bloch observed, no less speculative than those invoked by preformationists or creationists.[67]

To be fair, the figure of the modern Spinozist as an inventor of a new, more empiricist approach to transformations in the sciences can be found elsewhere in this cultural context in the eighteenth century—that is, it did not spring fully formed from Vernière's imagination—for instance in the Abbé Lelarge de Lignac's three-volume work against fatalism *Le témoignage du sens intime et de l'expérience opposé à la foi profane et ridicule des fatalistes modernes*. In this work, which appeared in 1760 (thus five years prior to the above *Encyclopédie* article), Lignac describes what he terms a "new Spinozism":

> Our Spinozists are quite subtle reasoners. They abandon the materialism of Locke and revise other points of the doctrine of this Englishman to which they are strongly attached. They mingle adroitly with those of our authors who... are drunk with the purported beauty of Locke's Metaphysics and tend towards universal Spiritualism. Our scattered (*déliés*) Spinozists *completely give up on the method of reasoning by abstraction and consequently on Spinoza's method*—they decline to recognise him as their master, and thus are always angry when they are called

[66] Namely, if Haller's physiology contributed the idea of a combinatorial system composed of the structural elements of the organism, a system of functional vital properties expressed at various levels of organic integration, Diderot in contrast is either (a) just a commentator on such concepts, (b) a materialist philosopher seeking to accumulate information to support his general metaphysics, or (c), more creatively, a thinker whose reflections on sensibility and fibres, organs, brains, bodies and networks constitute a genuine expansion of vitalist life science, boosted by speculative claims.

[67] Bloch, cited in Bourdin, "*Du Rêve de D'Alembert* aux *Éléments de physiologie*," 52–53.

> Spinozists.*But they claim to ground Spinoza's system in facts and deserve the glory of a second invention.*[68]

And there is plenty of reference to Spinoza and Spinozism in the life sciences in the period, usually in pejorative terms (just like the usage of those terms in philosophy), as in the accusation against Boerhaave that he was a Spinozist.[69] In addition, Vernière's category of neo-Spinozism is not so restrictive as to make its actors—here Maupertuis and Diderot—strictly a combination of Spinozism and Enlightenment life science; he also allows for the integration of Leibnizian components, which make much more sense of Diderot's way of understanding the self-organization of living matter, which at times takes the form of a "materialization of the monad."[70] But at the level of a *framework*, Diderot chooses to call it Spinozist, and instead of disqualifying this automatically, we can also consider it either a kind of performative rather than textually strict Spinozism, or a more 'constructivist,' home-grown species of the theory, what Ann Thomson once called "Spinosism with an 's' rather than a 'z,'" referring to the spelling of Diderot's article.[71] Unlike the more common case, when it is the danger of Spinozism that leads thinkers to invent new conceptual tools (from Cudworth and Goclenius to Kant and Fichte[72]), here it is a positively endorsed form of Spinozism that is invented.

[68] Lelarge de Lignac, *Le témoignage du sens intime et de l'expérience opposé à la foi profane et ridicule des fatalistes modernes*, 3 vols. (Auxerre, FR: F. Fournier, 1760), vol. 1, 350–351, emphasis mine. Vernière notes (ibid.) that de Lignac was close to Réaumur and thus 'up to date' regarding biological research and the sorts of ideological claims relating to such research.

[69] The story of how Boerhaave was accused of Spinozism is well known and can be traced back to his funeral oration (I thank Theo Verbeek for this point). Haller speaks of Boerhaave's "carelessness," which led him, in an encounter with a cleric who was attacking Spinoza, to "defend the atheist and hurt the Christian" (letter to Rast of June 5, 1777, in *Correspondance inédite de Albert de Haller, Barthez, Tronchin, Tissot avec le Dr. Rast, de Lyon*, ed. Dr. Vernay [Lyon, FR: Aimé Vingtrinier, 1856], 29). Boerhaave cited Spinoza in his *Praelectiones academicae*, ed. Haller, 3 vols. (Göttingen, DE: Anton Vandenhoeck, 1739), §§570, 578; and La Mettrie associates Boerhaave with Spinoza in his *Abrégé des systèmes, Œuvres philosophiques*, ed. F. Markovits (Paris: Fayard, 1987), I, 267. In a strictly medical context, Boissier de Sauvages remarks that mental illness is not always "a bodily flaw, as Boerhaave implies and the Spinozists assert"; *Nosologie méthodique*, 10 vols. (Lyon, FR: Bruyset, 1772), VII, 19.

[70] See, e.g., the way Diderot moves from the conatus to the Leibnizian nisus, in his consideration of atoms in the article "Hobbisme" (*Encyclopédie* VIII, 235).

[71] Ann Thomson, "Les Lumières radicales sont-elles panthéistes?" in *Qu'est-ce que les Lumières 'radicales'? Libertinage, athéisme et spinozisme dans le tournant philosophique de l'âge classique*, eds. C. Secrétan, T. Dagron, and L. Bove (Paris: Éditions Amsterdam, 2007), 258. On the idea of an 'invented Spinozism,' see Yves Citton, *L'Envers de la liberté. L'invention d'un imaginaire spinoziste dans la France des Lumières* (Paris: Éditions Amsterdam, 2006).

[72] For an interesting claim that German Idealist metaphysics (from Goclenius and Wolff to Fichte and Hegel) emerges specifically as a response to the danger of Spinozism, see Pierre-François

4. Conclusion: Spinozism and/as Epigenesis

Diderot's Spinozist biology and epigenesis as the moment of invention of vital materialism are located in an unusual conceptual space—neither a standard contribution to metaphysical or otherwise systematic discussion of Spinoza, nor a contribution to positive life science destined to become a chapter in a history-of-science survey. One could then imagine that he chooses to describe "modern Spinozists" as partisans of epigenesis in order to underscore the radicality of this biological theory, which, as we saw in *Le Rêve de D'Alembert,* "can overturn all schools of theology." Spinozism as epigenesis is not, then, a contribution to positive life science, yet we should not lose sight of the characteristic biologism of the concept, noticeable for instance in its anti-mathematicism. That is, partly like the polyp, monsters, and spontaneous generation, epigenesis belongs to a kind of "folk biology" that cannot be properly subsumed by the history of biology as a science. And yet, without reverting to the positivistic overtones of Vernière's conception in which Diderot's Spinozism is "in agreement with scientific data," we might concede that modern Spinozists of this sort seek to reconstruct a metaphysics on physiological bases, as Alexandre Métraux has suggested,[73] but not as inductive generalization.

If, contrary to earlier readings fixated on the 'myth of the precursor,' Diderot belongs neither to the history of biology nor to its prehistory,[74] the question remains: How should one account for his complex relation to science, especially life science, which is neither metaphysically grounded (like Descartes's *arbor scientiae*) nor a strictly inductive project (like Bacon's tables of experiment)? Not only does his Spinozist biology not fit in the narratives proposed by either the history of biology or (internalist) history of philosophy; it furthermore also weakens the basis on which some ideologically motivated historians of materialism claim that philosophical materialism is the 'handmaiden' of the natural

Moreau, "Wolff et Goclenius," *Archives de philosophie,* vol. 65, n° 1 (2002): 7–14. On the English context, see Rosalie Colie, "Spinoza and the Early English Deists," *Journal of the History of Ideas,* vol. 20, n° 1 (1959): 23–46.

[73] Alexandre Métraux, "Über Denis Diderots physiologisch interpretierten Spinoza," 131.

[74] For an early attempt to present "Diderot's biology" (not Diderot as 'precursor of Darwin,' but as a biologist), see Ferdinand Paitre, *Diderot biologiste* (1904; reprint, Geneva, CH: Slatkine, 1971): "Le plus glorieux titre de Diderot à l'admiration de l'historien, c'est…d'avoir été le premier transformiste" (89). For an attempt (of the sort made more brilliantly by Canguilhem) to explain why it is wrong to present Diderot as an evolutionist or transformist (or a precursor thereof), see Lester G. Crocker, "Diderot and 18th-Century French Transformism," in *Forerunners of Darwin, 1745–1859,* eds. B. Glass, O. Temkin, and W. L. Straus (Baltimore, MD: Johns Hopkins University Press, 1959).

sciences.[75] As Olivier Bloch put it, science is not "the laboratory of materialism,"[76] or if it is, it is so in a very pluralistic sense.

Diderot's presentation of "modern Spinozism" as epigenesis plus "ancient Spinozism," then, is (i) a serious engagement with the life sciences in flux in the mid-eighteenth century (and an anti-mathematical one); (ii) a metaphysical project in which natural history and physiology are in the service of materialism (but a vital materialism of active, self-transforming matter); and (iii) a speculative project (with a 'radical' dimension) that as such does not belong to the history of science and cannot be subsumed within it.

Acknowledgment

An earlier version of this chapter was presented at the Early Modern Medicine and Natural Philosophy Conference, Center for Philosophy of Science, University of Pittsburgh, in November 2012. I am grateful to the participants for their comments.

[75] A recent reiteration of the old—inseparably Marxist and positivist—view in which materialism and science are allies supporting each other's struggles is P. Charbonnat, *Histoire des philosophies matérialistes* (Paris: Syllepse, 2007); for some critical commentary, see M. Aury, C.T. Wolfe, "Sommes-nous les héritiers des Lumières matérialistes?" *Phares* 8 (2008): 11–33; also online at http://www.ulaval.ca/phares/vol8-08/texte02.html.

[76] Olivier Bloch, "Poursuivre l'histoire du matérialisme," in *Matière à histoires* (Paris: Vrin, 1998), 459.

PART FOUR

THE ORDER OF THE LIVING WORLD

11

On the Continuity of Nature and the Uniqueness of Human Life in G. W. Leibniz

LEA F. SCHWEITZ

1. Introduction

In a letter from 1686, G. W. Leibniz (1646–1716) writes the following to Antoine Arnauld (1612–94): "The rational soul is created only at the time of the formation of its body, *being entirely different from the other souls we know*, because it is capable of reflection and it imitates the divine nature on a small scale."[1] Here, Leibniz claims that rational souls, with humans as the example we know best, are created when the body is formed and stands apart from the other souls we know. In the context of the emerging life sciences, Leibniz finds that human life

[1] G, ii: 75 (*The Leibniz-Arnauld Correspondence*, M 93, emphasis added). Abbreviations are as follows: A: *G. W. Leibniz: Sämtliche Schriften und Briefe*, ed. Akademie der Wissenschaftern (Berlin: Akademie Verlag, 1923); cited by series, volume, and page; AG: *G. W. Leibniz: Philosophical Essays*, eds. and trans. Roger Ariew and Daniel Garber (Indianapolis, IN: Hackett, 1989); AT: *Oeuvres de Descartes*, 12 vols., eds. C. Adam and P. Tannery (Paris: Vrin/CNRS, 1964–76); cited by volume and page; CSM: *The Philosophical Writings of Descartes*, 2 vols., eds. and trans. J. Cottingham, R. Stoothoff, and D. Murdoch (Cambridge, UK: Cambridge University Press, 1988); cited by volume and page; G: *Die Philosophichen Schriften von G. W. Leibniz*, ed. C. I. Gerhardt (Berlin: Weidmann, 1875–90; reprinted Hildesheim, DE: Olms, 1978); cited by volume and page; L: *G. W. Leibniz: Philosophical Papers and Letters*, ed. and trans. Leroy E. Loemker, 2nd ed. (Dordrecht, NL: Reidel, 1969); LS: G. W. Leibniz, *Leibniz: The Shorter Leibniz Texts*, ed. and trans. Lloyd Strickland (New York: Continuum, 2006); M: *The Leibniz-Arnauld Correspondence*, ed. and trans. H. T. Mason (Manchester, UK: Manchester University Press, 1967); NE: *G. W. Leibniz: New Essays on Human Understanding*, eds. Peter Remnant and Jonathan Bennett (Cambridge, UK: Cambridge University Press, 1982); cited by book, chapter, and page; and WF: *Gottfried Wilhelm Leibniz: Leibniz's 'New System' and Contemporary Associated Texts*, eds. and trans. R. S. Woolhouse and Richard Francks (Oxford, UK: Oxford University Press, 1997).

is something special; humans reflect and imitate God—they are divine mirrors. At the same time, the formation of the body is not discounted. Humans are not only divine mirrors, but like all other living organisms, they are natural machines and may be classified into species that are determined by lines of generation.

One of the challenges of Leibniz's view of natural machines is how it coheres with his view of intelligent substances as divine mirrors, particularly in light of his other commitments to minimize God's miraculous activity and to maintain the law of continuity. Despite the challenges to a strong sense of divine mirroring, Leibniz seems to affirm it. In this chapter, I argue that Leibniz has the resources to meet these challenges; however, it demands re-envisioning his view of nature. The first part of the chapter develops Leibniz's view of nature as full of natural machines and concludes with the challenges such a view raises for human life. In the second part of the chapter, I outline Leibniz's view of nature as sacramental. I argue that his views on the infinite structure of natural machines, the *imago dei*, and omnipresence suggest that Leibniz sees the natural order as a place where God is immanent in finite things. The chapter concludes by arguing that Leibniz's sacramental view of nature can help address the challenges raised by his understanding of human life as both a natural machine and a divine mirror. In the end, I suggest that while living organisms may be classified into species defined by lines of generation, or relationships between parents (or potential parents) and offspring, Leibniz's view of human life is classified distinctively by its relationship to God.

2. Leibniz's Anthropology: Humans as Natural Machines and Divine Mirrors

In understanding the workings of the human body, Leibniz saw René Descartes's philosophy as "the entrance hall to the truth."[2] Cartesian bodies are just the "objects of geometry made real."[3] As a result, all the functions of bodies may be explained in terms of geometrical properties, such as size, shape, and motion. The view that emerges further results in the human body being likened to a machine:

> Indeed, one may compare the nerves of the machine I am describing [namely, the human body] with the pipes in the works of these fountains, its muscles and tendons with the various devices and springs which serve to set them in motion, its animal spirits with the water

[2] G, III: 606 (*Letters to Nicolas Remond*, L, 655).

[3] Daniel Garber, "Leibniz: Physics and Philosophy," in *Cambridge Companion to Leibniz*, ed. Nicholas Jolley (Cambridge,UK: Cambridge University Press, 1995), 284.

> which drives them, the heart with the source of the water, and the cavities of the brain with the storage tanks.[4]

Particularly evident in the *Treatise on Man* is Descartes's attempt to explain the workings of the body without recourse to the vegetative or sensitive souls of the scholastics. Descartes (1596–1650) takes the body "to be nothing but a statue or machine made of earth," and it contains within itself, as extended matter alone, everything it needs to perform a whole array of functions.[5]

To contemporary ears, this may not give much pause, but Descartes was proposing something quite remarkable. He was trying to wrest explanations of bodily functions away from longstanding scholastic models. By the end of the *Treatise on Man*, the list of functions Descartes claimed to ascribe to the bodily machine included digestion, the beating of the heart, growth, respiration, waking, sleeping, the function of the sense organs and memory, the movements of the appetites, the passions, and the limbs.[6] The details of Descartes's descriptions of these functions need not enter the discussion here; it is his attempt to mechanize the human body that is relevant for understanding Leibniz's anthropology. Descartes writes:

> I should like you to consider that these functions follow from the mere arrangement of the machine's organs every bit as naturally as the movements of a clock or of the automaton follow from the arrangement of its counter-weights and wheels. In order to explain these functions, then, it is not necessary to conceive of this machine as having any vegetative or sensitive soul or other principle of movement and life, apart from its blood and its spirits, which are agitated by the heat of the fire burning continuously in its heart—a fire which has the same nature as all the fires that occur in inanimate bodies.[7]

For Descartes, blood and spirits are sufficient to explain all the functions of the body, including movements and life: "It is not necessary to conceive of this machine as having any other principle of movement or life, apart from its blood

[4] AT, XI: 131 (*Treatise on Man*, CSM, I: 100). Animal spirits here should not be understood as spiritual per se. Rather, they are very fine particles that are separated from blood and are fast-moving because they retain heat from the heart. These particles move from the heart to the brain and eventually into the nerves. Though small, they are still fundamentally corporeal and part of the bodily machine.

[5] AT, XI: 120 (*Treatise on Man*, CSM, I: 99).

[6] AT, XI: 201–202 (*Treatise on Man*, CSM, I: 108).

[7] AT, XI: 202 (*Treatise on Man*, CSM, I: 108).

and its spirits," which are also explainable by natural, mechanical heat. Matter in motion explains Cartesian bodies and their functions.

Insofar as the human body is an extended machine, Leibniz is completely in agreement with Descartes's comprehensive mechanism. In the *Discourse on metaphysics*, Leibniz is unequivocal about the matter: "All particular phenomena of nature can be explained mathematically or mechanically by those who understand them."[8] Leibniz was quite optimistic about the power and reach of mechanical explanation and concurred with the moderns that everything in nature could be explained under the mechanical project. However, Descartes's mechanism is only the entrance hall to the truth; for Descartes, human bodies are simply artificial machines, but for Leibniz, human bodies are complex natural machines. According to Leibniz, Descartes's mechanization reduces natural things to artificial machines. This is problematic because it engenders a theological insult by anthropomorphizing God's creative ability. Leibniz claims that God's machines are nothing like human machines, and without a sufficiently rich view of nature, we risk confusing God's creative capacities with our own. Leibniz finds that Descartes has made precisely this mistake:

> I am as ready as anyone to do justice to the moderns; nevertheless I think they have carried reform too far, among other things in conflating natural things with artificial ones, through not having sufficiently grand ideas of the majesty of nature. They take the difference between nature's machines and ours to be only that between great and small. . . . I think that this gives an inappropriate and unworthy idea of nature, and that it is only my system which shows the true, and immense distance there is between the least productions and mechanisms of divine wisdom and the greatest masterpieces produced by the skill of a limited mind—a difference which is not merely one of degree, but one of kind.[9]

According to Leibniz, there ought to be no comparison between God's machines and ours. Justice to the Creator requires a difference between natural and artificial machines that is not merely a difference in degree, but rather, it must be a difference in kind. Cartesian matter only allows for a difference that is between "great and small," and this, according to Leibniz, minimizes the grandeur of nature and is unworthy of the Creator. Injustice to the Creator explains why limiting nature to artificial machines is problematic, but it does not yet explain what makes artificial machines so limited. What is the difference between natural and artificial machines for Leibniz?

[8] A, VI.iv: 1559 (*Discourse on Metaphysics*, AG, 51).
[9] G, IV: 481–482 (*New System of the Nature of Substances*, WF, 15–16).

The relevant difference, here, is not between machines and non-machines. It is not that artificial machines are simply machines and natural "machines" are something else. Leibniz is satisfied with Descartes's comprehensive mechanization project: "Everything in nature happens mechanically." The difference is one of complexity. Both Descartes and Leibniz agree that matter in motion can explain bodies, and they also agree that the matter that explains bodies is infinitely divisible. It is not that Cartesian artificial machines are not complex; they consist of aggregates of matter, which can be divided without end. However, they are not infinitely complex, and this marks the difference between a natural and an artificial machine. Artificial machines are infinitely divisible, but natural machines are infinitely divisible and infinitely complex. Leibniz thinks that this is the basis of the difference in kind between nature's machines and ours.

Leibniz builds infinite complexity into his view of nature through the nested structure of living organisms: "Nature's machines have a truly infinite number of organic parts."[10] Descartes's artificial machines have an infinite number of material parts because they are infinitely divisible, but Leibniz's natural machines have an infinite number of organic parts because they are infinitely nested with other organic bodies.[11] The infinite complexity that grounds the difference between natural and artificial machines requires substantial forms because they provide the unifying structure that allows for the nested or branching structure inherent in Leibniz's infinitely complex natural machines.[12] Although the forms provide the unity that is necessary for corporeal substances, infinite complexity is the distinguishing characteristic of natural machines.

Leibniz illustrates the difference to Arnauld by considering the difference between a corpse and a living person. A human corpse amounts to an artificial machine. It is infinitely divisible, but it no longer has the infinite complexity required for being a natural machine or the true unity required for being a corporeal substance. Leibniz writes to Arnauld: "In my opinion, our body in itself or the *cadaver*, setting the soul apart, can be called a substance only in an improper sense, just as in the case of a machine or a pile of stones, which are only beings by aggregation; for regular or irregular arrangement does not

[10] G, IV: 482 (*New System of the Nature of Substances*, WF, 16).

[11] This account of natural machines follows on the fine work of Michel Finchant, "Les machines de la nature," *Studia Leibnitiana* 35.1 (2003): 1–28; Ohad Nachtomy, "Leibniz on Artificial and Natural Machines: Or What it Means to Remain a Machine to the Least of its Parts," in *Machines of Nature and Corporeal Substances in Leibniz*, eds. Justin E. H. Smith and Ohad Nachtomy (Dordrecht, NL: Springer, 2010): 61–80; and Justin E. H. Smith, *Divine Machines: Leibniz and the Sciences of Life* (Princeton, NJ: Princeton University Press, 2011).

[12] For further development of Leibniz's nested individuality, see Ohad Nachtomy, "Leibniz on Nested Individuals," *British Journal for the History of Philosophy*, 15.4 (2007): 709–728; and his *Possibility, Agency, and Individuality in Leibniz's Metaphysics* (Dordrecht, NL: Springer, 2007), esp. chapter 9, 215–235.

constitute substantial unity."[13] We can see here that Leibniz takes humans to be substances in the proper sense, but only insofar as we are natural machines whose souls create a substantial unity. With respect to Leibniz's anthropology, this means that humans cannot be just any kind of machine, as Descartes seems to say in comparing the human with a fountain. For Leibniz, humans (and all other living things) are natural machines—animate machines rather than artificial machines. Leibniz thought he had experimental science on his side. When the first microscopists cut up and peered into living things, they found more living things inside. Leibniz saw no prima facie reason why this pattern could not continue ad infinitum.[14]

We are now in a position to understand the full import of the explicit description of a human being that Leibniz gives to Arnauld in 1687.[15] He writes that the human is "an entity endowed with a genuine unity conferred on him by his soul, notwithstanding the fact that the mass of his body is divided into organs, vessels, humours, spirits, and that the parts are undoubtedly full of an infinite number of other corporeal substances endowed with their own entelechies."[16] Humans are pictured as complex, living entities with infinitely complex parts that are also full of complex, living entities with complex parts. This understanding of the human body allows Leibniz to maintain both the infinite divisibility of matter and the mechanization of the human body while providing a ground for the unity that is the basis for the infinite complexity characteristic of a natural machine. Considered in itself, the human body is an artificial machine. However, the living human being is an infinitely complex natural machine containing an infinite number of organic parts. This difference marks a difference in kind; it does not confuse our artificial machines with God's natural machines, and the position is not subject to the charge of committing a theological insult against God, the Creator, who chose these creatures as the expression of the best of all possible worlds.

The category of natural machines allows Leibniz to carve the created world up into the living and the nonliving—but not along lines that common sense

[13] G, II: 75 (*The Leibniz-Arnauld Correspondence*, AG, 78).

[14] For more on the impact of the microscope in the early modern period, see Catherine Wilson, *The Invisible World: Early Modern Philosophy and the Invention of the Microscope* (Princeton, NJ: Princeton University Press, 1995).

[15] This chapter focuses on Leibniz's middle to later writings, but this is not meant to suggest that these views represent Leibniz's final position. The development of Leibniz's thought on these themes is worth exploring, but it is beyond the scope of the current chapter. For a thorough treatment of Leibniz's early years, see Christia Mercer, *Leibniz's Metaphysics: Its Origins and Development* (Cambridge, UK: Cambridge University Press, 2001). For a recent treatment of the development of Leibniz's thought, see Daniel Garber, *Leibniz: Body, Substance, Monad* (Oxford, UK: Oxford University Press, 2009).

[16] G, II: 120 (*The Leibniz-Arnauld Correspondence*, M, 154).

might suggest. Nature is full of living things for Leibniz: The created order is literally full of infinitely divisible matter and infinitely complex organic bodies, which are never fully separated from souls.[17] As Smith concludes, "Every *corps* is the bodily component of some corporeal substance, and all corporeal substances are in the end what we would think of as biological entities."[18] On such a view, it is not clear what to make of human uniqueness. Human life seems to be characterized as biological entity, no different than any other organic body or natural machine.

Leibniz worries about classification as well. Smith argues that Leibniz addresses this worry with the reproductive species concept.[19] Reproductive species are domain-specific; the concept does not apply to all the objects of Leibniz's concern. Notably absent are mathematical objects, such as triangles or ellipses, and inanimate objects, such as gold or clocks. The *Nouveaux essais* give attention to the identity conditions of these inanimate things and the ways in which our language captures them. However, Smith persuasively argues that Leibniz also works with a reproductive species concept that applies specifically to the domain of the living. Smith's view of Leibniz's reproductive species concept maps the domain of the living with real species defined by lines of generation. A species through generation includes living organisms that share or could share an origin. For Leibniz, organic bodies are classified in species that are defined by generation, that is, by consideration of the lineage of the offspring.[20]

Before the increasingly mechanical understandings of reproduction took hold, offspring were thought to receive a form that provided an individual's real essence and grounded its species membership. As such, the mechanization of embryogenesis threatened to erode the world of species and leave only individuals. Individuals were produced, but nothing in them was reproduced in a way that would ground species membership or cluster natural organisms into natural kinds.[21] Without a form to direct nature's processes and ground the reality of

[17] Smith follows Donald Rutherford in claiming that this interpretation is consistent with Leibnizian monadological immaterialism. See Rutherford, *Leibniz and the Rational Order* (Cambridge, UK: Cambridge University Press, 1995), 230; and Justin E. H. Smith, *Divine Machines: Leibniz and the Sciences of Life* (Princeton, NJ: Princeton University Press, 2011), 124. For the purposes of this chapter, I assume this without argument. The subtle debate surrounding Leibniz's idealism is beyond the scope of this chapter.

[18] Smith, *Divine Machines*, 124–125.

[19] Smith calls this the biological species view. Although he recognizes that this is anachronistic, he keeps the term. I find that reproductive species more accurately represents Leibniz's view of species by generation. Not only does it pick out the key feature of the concept, but it also provides some distance from the contemporary biological species concept.

[20] For more on early modern views of animal generation and reproduction, see Justin E. H. Smith, ed., *The Problem of Animal Generation in Early Modern Philosophy* (Cambridge, UK: Cambridge University Press, 2007).

[21] Smith provides examples in the work of Nicolas Malebranche and Nathaniel Highmore. See Smith, *Divine Machines*, 204–206; 239–240, respectively.

species, classifications of the natural order are based on nothing more than our own pragmatic (and relatively arbitrary) decisions.[22] However, Leibniz is able to maintain the reality of reproductive species: "In contrast to mathematical identities, and in contrast with ordinary physical entities, the criteria for establishing sameness, or sufficient sameness to be included in the same species, are not based on arbitrary or subjective concerns of the person who classifies living beings but on the objective fact that all the members of the same species of living beings have the same origin"; this fact is independent of human construction and may ground an objective, non-arbitrary species.[23] On this basis, reproductive species are real species and may be used to classify living things.

In addition, reproductive species are one of the ways in which Leibniz understands plants and animals as divine machines—God chose these living things to create the best of all possible worlds, preformed them with naturally immortal souls, and endowed them with a capacity for self-reproduction. As such, each divine machine bears real relations to other members of its species in virtue of the similarities in their generative systems.[24] Reproductive species share origins; species by generation are determined by descent from parents that share an origin and lineage. As a result, morphological differences do not impact membership in reproductive species.[25] In the end, physical resemblances are not determinative of reproductive species (even if they may provisionally ground physical species). The reproductive species concept is necessary for Leibniz's understanding of the human species; like all other living organisms, human life can be classified into real species as defined by lines of generation. However, it is not sufficient to capture what Leibniz seems to take as definitive of intelligent substances, such as human beings. Leibniz characterizes human life not only as a natural machine defined by its origins, but also as a divine mirror defined by its relationship to God.

However, the challenge is as follows: If the human species is a reproductive species unfolding in a constitutive relationship with its parents, there does not appear to be room or much need for God's involvement. Furthermore, any talk about God's involvement in a nature that is understood as above seems to demand God's miraculous intervention—something which Leibniz sought to minimize whenever possible. Such a concern was at the heart of Leibniz's critiques of Nicolas Malebranche's occasionalism; he worries that divine interventions are a

[22] Leibniz's strict identity requirements, as described in his complete individual concept and resulting logical species, make this dynamic acute, and it is a repeated locus of concern in his *New Essays on Human Understanding*.

[23] Smith, *Divine Machines*, 247.

[24] In this, Smith finds a foreshadowing of Kant's "unity of generative powers." See Smith, *Divine Machines*, 246.

[25] Smith, *Divine Machines*, 251.

challenge to the divine wisdom that is expressed in the order and beauty of the creation. Yet, in Leibniz's writings, we find that the commitment to the uniqueness of rational souls seems to push him in that direction. However, there is logical space for Leibniz to maintain a view of humans as divine mirrors (defined by a relationship to God) that does not commit Leibniz to a miraculous intervention of God. In the next section, I outline Leibniz's sacramental view of nature by drawing together several lines of current scholarship that take seriously God's immanence in the natural order. From the context of a sacramental view of nature, Leibniz has resources to affirm the uniqueness of intelligent life, resist the need for God's miraculous intervention, and maintain the continuity of the natural order.

3. Leibniz's Sacramental View of Nature

Leibniz repeatedly claims that intelligent substances are divine mirrors: "As for the rational soul, or mind, there is something more in it than in monads, or even simple souls. It is not only a mirror of the universe of created things, but also an image of the divinity."[26] As a divine mirror, humans have an ongoing relationship with God. Mirroring may be understood in a strong ontological sense in which intelligent substances reflect and are constituted by the divine light within, or in a weaker analogical sense in which intelligent substances are somehow independent expressions of the divine light from without. Both senses are potentially problematic. The weaker sense either elides a genuine, active relationship in the mirroring, or it depends upon God's miraculous intervening involvement into the creation. The stronger sense seems to push against some of the foundations of Leibniz's philosophical system. For example, it is crucial for Leibniz that the created order has genuine agents who are able to produce their own effects.[27] If mirroring is understood in the strong ontological sense, the boundaries between Creator and creation seem to blur in ways that compromise creaturely autonomy and spontaneity. In this chapter, I hope to simply suggest three aspects of Leibniz's thought that speak in favor of the strong ontological sense of mirroring. Leibniz's views on the infinite complexity of nested individuality, an expanded sense of the *imago dei*, and God's omnipresence recommend a deep relationship between God and the creation—despite the challenges that seem to arise from his core philosophical principles. Within the constraints of the present chapter,

[26] G, VI: 604 (*Principles of Nature and Grace*, AG, 211–212).

[27] For a nuanced account of spontaneity and its importance in Leibniz's thought, see Donald Rutherford, "Leibniz on Spontaneity," in *Leibniz: Nature and Freedom*, eds. Donald Rutherford and J. A. Cover (Oxford, UK: Oxford University Press, 2005), 156–180.

I can only hope to motivate further research into whether Leibniz was in fact able to hold such views consistently and to what degree they open him up to charges of Spinozism. However, a recent work on Leibniz's views on divine concurrence and divine conservation that attends to similar challenges concludes that "Leibniz's metaphysical and theological commitments can be neatly woven together."[28]

In this section, Leibniz's views on nested individuality, the *imago dei*, and divine omnipresence will be outlined as examples of nature's sacramental structure. Nature is neither bereft of God's presence and activity, nor is it equivalent to it. A sacramental view of nature sees the whole of the created order as exhibiting one of the key principles of Lutheran sacramental theology, namely, that the finite is capable of the infinite (*finitum capax infiniti*).[29] Lutheran sacramental theology affirms that finite matter in the forms of bread, wine, and water is a means of grace and a vehicle for the divine.[30] Said another way, the sacraments are instances when the "finite is capable of the infinite." The material elements of the sacraments are means of real and transformative encounters with the divine because they are capable of the infinite in, with, and under the finite.

We have already seen one instance of this view of nature in the infinite complexity of natural machines. The nested individuality of living things captures one aspect of the interplay between the infinite and the finite in Leibniz's view of nature. A natural machine is a machine each of whose parts are machines and whose nested structure is characterized by infinite complexity. The curious feature of Leibniz's thought that makes Nachtomy's model of nested individuality relevant for a sacramental view of nature is the ubiquity of living beings in Leibniz's view. Nature for Leibniz is replete with living things understood on the metaphysical model of nestedness; one might go so far as to say that the priority in Leibniz's view of nature is on the great chain of the living. As a result, nested individuality describes a uniform theory of nature that is teeming with life. On this view, living things are not simply finite expressions or pale reflections of the infinite. Constitutive of the definition of a living thing is its infinite

[28] Jeffrey McDonough, "Leibniz: Creation and Conservation and Concurrence," *Leibniz Review*, 17 (2007): 55.

[29] For an explicit treatment of Leibniz's Lutheran heritage, see Ursula Goldenbaum, "Leibniz as a Lutheran," in *Leibniz, Mysticism, and Religion*, eds. Allison Courdert, Richard H. Popkin, and Gordon M. Weiner (Dordrecht, NL: Kluwer, 1998), 169–192. In addition, relevant considerations throughout Leibniz's life may be found in Maria Rosa Antognazza, *Leibniz: An Intellectual Biography* (New York: Cambridge University Press, 2009).

[30] The "Augsburg Confession," e.g., states: "Concerning the Lord's Supper it is taught that the true body and blood of Christ are truly present under the form of bread and wine in the Lord's Supper and are distributed and received there." Robert Kolb and Timothy J. Wengert, eds., "Augsburg Confession," in *The Book of Concord: The Confessions of the Evangelical Lutheran Church*, trans. Charles Arand, et al. (Minneapolis, MN: Augsburg Fortress Press, 2000), 44.

structure, and living things are everywhere. Living things are capable of the infinite in the strong sense: They are not only finite expressions of God's infinite power and perfection; they are infinite in important respects.[31] Absolute infinity may be reserved for God, but living things participate (in finite, particular ways) in God's infinity.[32]

The sacramental feature of Leibniz's view of nature can be seen in his expanded understanding of the doctrine of the *imago dei*, as well. Traditionally, the *imago dei* is understood as naming a characteristic of human beings: Humans, in light of some distinctive feature, are so closely related to God as to be made in God's image. Whether the likeness is literal or figurative is a matter of debate, as is the purported distinctive feature—although rationality is usually a top contender. Traditionally, the doctrine limits the extension of the class of creatures for whom the image of God applies to human beings (or, sometimes, intelligent substances).

Allan Hillman argues that Leibniz understood all created substances to be in the likeness of God insofar as they share God's causal powers or perfections. In this sense and contrary to some traditional views, the *imago dei* extends to all living creatures. In the *Discourse on Metaphysics*, we find a powerful statement of this expanded understanding of the *imago dei*:

> It can even be said that every substance bears in some way the character of God's infinite wisdom and omnipotence and imitates him as much as it is capable. For it expresses, however confusedly, everything that happens in the universe, whether past, present or future—this has some resemblance to an infinite perception or knowledge. And since all other substances in turn express this substance and accommodate themselves to it, one can say that it extends its power over all the others, in imitation of the creator's omnipotence.[33]

On such a reading, every substance shares in God's wisdom and power; as such, the *imago dei* is extended to a wide range of created substances.[34] In this respect, created substances, *all* created substances, seem to share in God's infinite wisdom

[31] Thanks to Ohad Nachtomy for helpful suggestions in clarifying the point here.

[32] Nachtomy's analysis of infinity is relevant to this discussion and worth exploring further. See his contribution to this volume and Nachtomy, "Tale of Two Thinkers, One Meeting, and Three Degrees of Infinity: Leibniz and Spinoza (1675–8)," *British Journal for the History of Philosophy* 19.5 (2011): 935–961.

[33] A, VI.iv: 1552 (*Discourse on Metaphysics*, AG, 42).

[34] My reading follows Hillman's helpful analysis. His account connects the doctrine of *imago dei* to Aquinas and analyzes the implications for Leibniz's theory of containment and doctrine of creation. See T. Allan Hillman, "Leibniz on the *Imago Dei*," in *Oxford Studies in Early Modern Philosophy*, eds. Daniel Garber and Stephen Nadler (Oxford, UK: Oxford University Press, 2010), 171–212.

and omnipotence. It remains a finite imitation because of the particular ways in which created substances express God's infinity.[35] In this relationship between the finite and the infinite, the sacramental character of the expanded view of the *imago dei* is revealed.

Created substances are intimately related to God, so intimately, in fact, as to share in the same kind of perfection or power, though in lesser degree. They are finite, and they are something more. They are an expression of God's infinite perfections. Consider the following from a letter written in 1698 to André Morell:

> As all minds are unities, it can be said that God is the primitive unity, expressed by all the others according to their capacity. His goodness moved him to act, and there are in him three primacies: power, knowledge and will; the result of these is the operation or creature, which is varied according to the different combinations of unity and zero; or rather of the positive with the privative, for the privative is nothing other than limits, and there are limits everywhere in a creature, just as there are points everywhere in a line. However, a creature is something more than limits, because it has received some perfection or power from God, just as the line is something more than the negation of the progress beyond which it ends.[36]

In this passage, the finite is more than simply finite because it receives some of God's power or perfection and in so doing takes on the infinite.[37] Through the perfections received from God, created substances are quite literally made in the image of God or a kind of extension of God.[38] This expanded sense of the *imago dei* reveals the strongly sacramental structure of Leibniz's view of nature, wherein the finite is capable of the infinite.

Finally, Leibniz's view of God's omnipresence also suggests a sacramental view of nature. God is naturally present in every place; there is no "place" that created things can be where God is absent:

> I should think that everything is in God not as a part is in a whole, nor as an accident is in a subject, but as a place is in that which contains it,

[35] This is not to say that God's powers are completely infinite. Even God's power extends only as far as the logically possible. See Hillman, "Leibniz on the *Imago Dei*," 207–208.

[36] A, I.iii: 660 (*Letter to André Morell*, LS, 39).

[37] Robert Adams acknowledges that unresolved interpretative challenges remain in holding together Leibniz's view of divine perfection, his positive characterization of perfections, and the relation between limited and unlimited degrees of a quality. See Adams, *Leibniz: Determinist, Theist, Idealist* (New York: Oxford University Press, 1994), 115–119.

[38] Hillman notes that his view has some affinity to Christia Mercer's account, but it stops short of a Platonic emanation account. See his discussion in Hillman, "Leibniz on the *Imago Dei*," 205, fn. 76.

> a spiritual or a sustaining place.... God is immense or everywhere, and the world is present to him. And so, everything is in him, for he is where things are and where they aren't, and he remains when they depart, and is already there when they arrive.[39]

On the whole, Leibniz's view of God's omnipresence does not receive much attention because it is collapsed into consideration of God's operation and omnipotence. Given that Leibniz reduces all presence of any substance in a place to operation in that place, by extension, Leibniz "seems favorably disposed to the Scholastic reduction of God's presence in a place to 'immediate operation' on things that are in that place."[40] For our purposes, however, even this reduction of omnipresence to immediate operation demands that the infinite operates immediately on finite things, and as a result, exhibits a sacramental structure of nature.

4. Human Life and the Continuity of Nature

In the final section of this chapter, we return to the question of the continuity of nature and the uniqueness of human life, characterized as a natural machine and a divine mirror. From the perspective of Leibniz's sacramental view of nature, we can look anew at passages like the one that opened this chapter. There, Leibniz identifies the uniqueness of human life in terms of some feature of the rational soul, such as reflection, as the relevant difference between rational souls and the other souls we know. In the *Monadology*, Leibniz cashes out the difference as a peculiarity of human origins:

> As for minds or rational souls, I find that, at bottom, what we just said holds for all living beings and animals, namely that animals and souls begin only with the world and do not end any more than the world does. However, rational animals have this peculiarity, that their little spermatic animals, as long as they only remain in this state, have only ordinary or sensitive souls. But that as soon as the Elect among them, so to speak, attain human nature by actual conception, their sensitive souls are elevated to the rank of reason and to the prerogative of minds.[41]

Human beings have immortal souls that cannot be created or annihilated in the natural course of time; they are classified by reproductive species; and they are

[39] *Comments on Spinoza's Philosophy*, AG, 276.
[40] Adams, *Leibniz*, 124.
[41] G, VI: 621 (*The Monadology*, AG, 223).

infinitely complex, nested structures. In these ways, humans are like any other living organism. Yet, something happens to elevate rational souls to the rank of the "Elect." Natural machines and reproductive species alone seem unable to account for this. Not surprisingly, recent scholars have been concerned that this elevation of sensitive souls to the rank of reason occurs by means of God's miraculous intervention—which surely introduces a discontinuity into the natural order, among other problems. In the end, it seems that neither Leibniz scholars nor Leibniz himself are able to come down definitively on whether the elevation of the rational soul is through a miraculous divine intervention, a non-mirculous divine intervention, or via the soul's natural spontaneity.[42] Over the course of Leibniz's writings, the texts are genuinely ambiguous; the sacramental view of nature cannot overcome some of these inconsistencies in Leibniz's corpus. Rather, it simply tries to open up a third way between divine intervention and natural spontaneity. On this view, the interplay of the finite and the infinite embeds a measure of God's absolute infinity within the natural order. As such, divine activity need not be intervening, and nature's spontaneity need not wholly exclude the divine. Unfortunately, even with this revision, when it comes to the elevation of the rational soul, Leibniz's texts simply do not provide a definitive position. However, this is not the only characteristic that Leibniz signals as distinctive for human life.

In the *Discourse on Metaphysics*, the difference is cast another way. Here Leibniz writes, as follows:

> The difference between intelligent substances and substances that have no intelligence at all is just as great as the difference between a mirror and someone who sees. Since God himself is the greatest and wisest of all minds, it is easy to judge that the beings with whom he can, so to speak, enter into conversation, and even into a society—by communicating to them his views and will in a particular manner and in such a way that they can know and love their benefactor—must be infinitely nearer to him than all other things, which can only pass for the instruments of minds.[43]

These passages leave the reader convinced that Leibniz took there to be a difference in kind between intelligent, rational souls and every other mind. However, it is a difference marked by a relation between God and those with whom God can enter into a society; it is a relation that makes intelligent creatures infinitely

[42] See Daniel C. Fouke, "Spontaneity and the Generation of Rational Beings in Leibniz's Theory of Biological Reproduction," *Journal of the History of Philosophy*, 29.1 (Jan. 1991): 33–45; and Gregory Brown, "Miracles in the Best of All Possible Worlds: Leibniz's Dilemma and Leibniz's Razor," *History of Philosophy Quarterly*, 12.1 (Jan. 1995): 19–39.

[43] A, VI.iv: 1585 (*Discourse on Metaphysics*, AG, 66).

nearer to God than all other things. On this characterization of the human situation, the sacramental view of nature complements a created world of natural machines and suggests a way to resolve the tension between Leibniz's commitments to the principle of continuity and the uniqueness of human life.

It is not that humans or rational souls alone exhibit the sacramentality of nature; humanity is deeply embedded in a created order that resembles its Creator in the interplay between the finite and the infinite—and yet, there is a difference as great as the difference between a mirror and one who sees.[44] Species, as we have seen in the reproductive species concept, are non-arbitrary groups of individuals that are classified by a shared attribute or relation. Living organisms as natural machines may be classified by generation or shared reproductive lineages. Leibniz's consistent suggestion is that humanity's "Election" is to the city of God such that we must join morals to metaphysics in order to understand humanity's unique place in the natural world.[45] The *Monadology* continues:

> This city of God, this truly universal monarchy, is a moral world within the natural world, and the highest and most divine of God's works. The glory of God truly consists in this city, for he would have none of his greatness and goodness were it not known and admired by minds. It is also in relation to this divine city that God has goodness, properly speaking, whereas his wisdom and power are evident everywhere.... This harmony [between the physical kingdom of nature and the moral kingdom of grace] leads things to grace through the very paths of nature.[46]

As citizens in the city of God, human uniqueness emerges as a moral world within the natural world that is infinitely complex and sacramental in character. Continuity is maintained in two senses. The moral world that defines the human situation arises through the "very paths of nature," and miraculous intervention is unnecessary. The sacramental view of nature embeds natural machines with the infinite, thereby maintaining the continuity of nature even with God's relation.

Additionally, continuity is maintained through multiple continuous hierarchies. In the *New Essays*, Leibniz writes: "The Law of Continuity holds that nature leaves no gap in the orderings which she follows, but not every form or species belongs to each ordering."[47] Laurence Carlin draws on Leibniz's claim

[44] For additional evidence for this claim, consider the following two passages: (1) "Monads are 'as little gods made in the image of God and having in them some ray of the light of the Divinity'" (G, IV: 479; *New System*, L, 455); and (2) "For it can be said that created minds differ from God only in degree, from finite to infinite" (G, II:125; *Leibniz-Arnauld Correspondence*, M, 159–160).

[45] A, VI.iv: 1585 (*Discourse on Metaphysics*, AG, 66).

[46] G, VI: 621–622 (*The Monadology*, AG, 224).

[47] A, VI.vi: 307 (*New Essays Concerning Human Understanding*, NE, 307).

that "not every form or species belongs to each ordering" to offer an interpretation of nature as both continuous and discrete.[48] Whereas Leibniz's great chain of being applies to the totality of substances and the continuous ordering of metaphysical perfection, there are other types of perfections that form other continuous hierarchies. Humans, for example, belong to the continuous hierarchy of rational perfections, but not every species does. For Carlin, Leibniz's view of nature is continuous with respect to the metaphysical perfections of individuals and—at the same time—discrete with respect to other continuous hierarchies. Carlin explicitly includes the perfection of rationality, but the view could be extended to include reproductive species and moral perfections like citizenship. Recognizing multiple continuous hierarchies, Leibniz's view of nature is able to support a difference in kind that is compatible with a continuous metaphysical difference in degree. The advantage of the classification gained through a sacramental view of nature is that it characterizes intelligent substances in one of Leibniz's preferred categories—as citizens in the City of God. The classification organizes the natural world into discrete species that are based on moral perfections, and this classification depends on a non-miraculous, active relationship with God. Parallel to reproductive species that are defined by origins and a relationship to one's parental lineage, moral species are defined by ends and a relationship to God. More needs to be said to argue for the compatibility of the multiple, discrete species designations, but they come to the same conclusion: The natural order is both continuous and discrete.[49]

Leibniz's sacramental view of nature and the tensions surrounding his view of human uniqueness make three shifts visible. On the one hand, Leibniz is not a fully modern thinker. One of the great unresolved questions in contemporary evolutionary biology is "Why do organisms form discrete clusters instead of an organic continuum?"[50] This is not Leibniz's question. He does not interrogate why organisms are clustered in species instead of spread along a continuum, because Leibniz thinks organisms are clustered into species *and* spread along

[48] Laurence Carlin, "Leibniz's Great Chain of Being," *Studia Leibnitiana* 32.2 (2000): 131–165.

[49] Carlin is responding to objections raised by several scholars who conclude that the tension between the law of continuity and discrete differences of kind within the created order is unresolvable. Candice Goad and Susanna Goodin claim that there is no ready answer to these competing demands, and Donald Rutherford argues that "rationality introduces an infinite gap in Leibniz's continuous order of degrees." See Candice Goad and Susanna Goodin, "Monadic Hierarchies and the Great Chain of Being," *Studia Leibnitiana* 29.2 (1997): 145; and Donald Rutherford, *Leibniz and the Rational Order of Nature* (Cambridge, UK: Cambridge University Press, 1995): 165. Carlin's account helpfully suggests the possibility of multiple continuous hierarchies as previously described; in addition, he questions whether rational capacity alone is sufficient to open up a gap in the continuous order of nature. Both aspects of Carlin's response support the claims of this chapter. For more on rational capacities, see Carlin, "Leibniz's Great Chain of Being," 142–143.

[50] Jerry Coyne and H. Allen Orr, *Speciation* (Sunderland, MA: Sinauer Associates, 2004), 7.

a continuum. In these dual commitments, he faces a genuine tension, but his framework for the species problem with respect to the continuity of nature is not a problem of why nature is organized into species instead of a continuum, but rather how nature is both continuous and dense. For Leibniz, this tension is acute in the case of the human species, which he claims is both essentially different from other created organisms and part of the organic continuum. This is one version of the early modern form of the species problem. On the other hand, Leibniz is not fully a pre-modern thinker either. He is a contributor to the emerging life sciences with a mechanistic understanding of generation that grounds a classification of reproductive species. In the history of science, there are aspects of Leibniz's view that find a way into later understanding of species. The final shift is in the view of the created order itself. Throughout the chapter, I have used the terms human life, intelligent substances, and rational souls interchangeably, but this equivocation is not entirely accurate for Leibniz. For him, the class of intelligent substances extends above human beings to include higher creatures, like angels. My shorthand is excusable, however, because although Leibniz admits such higher creatures into the pantheon of the natural world, human beings are the intelligent substances we know best.

In Leibniz's view, humans live double lives as natural machines and divine mirrors. I have argued that Leibniz's sacramental view of nature helps to address the challenges to the principle of continuity that arise in his philosophical anthropology and supports a view of human uniqueness in terms of a constitutive relationship to God. Without such a view, commentators may be subject to a version of the charge Leibniz makes against Descartes. According to Leibniz, Descartes's view of the body confuses artificial machines and natural machines—and thereby insults God's creative activity. Similarly, Leibniz could argue that if we confuse mirrors of nature and mirrors of God, we insult God's goodness.[51] For Leibniz, human life is decidedly both a natural machine and a divine mirror; this chapter argues that Leibniz's sacramental view of nature can help make sense of the uniqueness of human life while maintaining the law of continuity—and thereby, it helps to avoid the confusion that results in theological insult.[52]

[51] When it comes to maintaining the proper place of humanity, Leibniz is rather serious. Consider, e.g., the following from the beginning of the *New Essays:* "I have read the famous Englishman's book [Locke's *Essay*] of which you have just spoken. I think very well of it, and have found fine things in it. But it seems to me that we should go deeper, and that we should even part company from his opinions when he adopts ones which limit us unduly" (A, VI.vi: 73; *New Essays on Human Understanding*, NE, I.i: 73).

[52] An early version of this essay was presented to the Early Modern Working Group at the Center for Philosophy of Religion, University of Notre Dame, and to the Leibniz Society of North America at the American Philosophical Association, Eastern Division meeting. Thank you to the participants, respondents, and organizers for very helpful suggestions, especially Christia Mercer, Ohad Nachtomy, and Sam Newlands.

12

Order of Insects

Insect Species and Metamorphosis between Renaissance and Enlightenment

BRIAN W. OGILVIE

1. Introduction

My title comes from a story by the American writer William Gass.[1] The unnamed narrator of "Order of Insects" has just moved into a new house with her husband and children. At first, she is distressed by the empty wing casings and molted skins of insects that each morning sully the floors of her otherwise immaculate house. As time passes, however, she comes to look forward to them and to devote more and more of her time to learning where they come from, their habits, and their classifications. What had been disgusting becomes familiar and then fascinating. With her family life increasingly fragile and disordered, she finds solace and perhaps strength in the chitinous regularity, the predictable habits—in short, the order—of insects.[2]

Europeans in the early modern period went through a similar shift, from revulsion to fascination, when thinking about insects. In particular, naturalists abandoned the notion that insects were produced through equivocal or spontaneous generation from decaying organic matter, an irregular process, replacing it with a concept of insects as subject to an orderly form of reproduction. But that was only one aspect of early modern Europeans' new interest in insects. In his groundbreaking two-volume *Materialien zur Geschichte der Entomologie bis Linné* (1928–29), F. S. Bodenheimer noted a sudden increase in European writing on insects

[1] "Order of Insects," in William Gass, *In the Heart of the Heart of the Country and Other Stories* (New York: Harper & Row, 1958).

[2] For a recent reading of the story, see Bertrand Gervais, "Reading as a Close Encounter of the Third Kind: An Experiment with Gass's 'Order of Insects,'" in *Insect Poetics*, ed. Eric C. Brown (Minneapolis and London: University of Minnesota Press, 2006), 223–245.

that began in the late 16th century and continued through the systematization of natural history by Carolus Linnaeus in the 18th. Particularly important were those scholars who were active in what Bodenheimer called "the bionomic period [*das bionomische Zeitalter*]," c. 1640–1750, when many of the laws of insect form and transformation were first set out.[3] Bodenheimer was concerned with natural science, but the entomologist Marcel Dicke has noted a concurrent increase in artistic representations of insects, in particular in Netherlandish still lifes.[4]

As early modern naturalists and artists turned their attention to insects, they confronted the problems of metamorphosis and diversity. Sixteenth-century naturalists, by and large, relegated these issues to second place, focusing instead on the synchronic morphology of insects. It was in the middle of the 17th century that scholars and artists problematized the successive transformations of insects from egg to imago (the adult, sexually mature insect), and around the same period that they began to pay serious attention to the immense diversity of the insect world.[5] The Dutch artist Johannes Goedaert devoted decades to observing and recording the transformations of insect larvae into pupas and imagos. The English naturalist Martin Lister found Goedaert's work fascinating, but he thought it needed to be organized and its descriptions classified before it could be truly useful. The German artist-naturalist Maria Sibylla Merian followed Goedaert's example, but with far more care and precision. Meanwhile, the Dutch anatomist Jan Swammerdam attempted, in his study of insect metamorphosis, to reduce the diversity of insects to four distinct classes based on modes of transformation.

Swammerdam's work, much of which lay unpublished until 1737–38, was the most thorough attempt in the early modern period to relate insect metamorphosis to classification. But in his focus on broad groupings, Swammerdam offered little help to naturalists who wanted to describe natural diversity. In their own ways, the Italian savant Antonio Vallisnieri and the French polymath René-Antoine Ferchault de Réaumur, in works published from the 1710s through the 1740s, attempted to develop a more pragmatic, anthropocentric (though not anthropomorphic) approach to ordering insects. Though aware of Swammerdam's work, they found it less than satisfactory for making sense of the myriad insect kinds that they, as field naturalists, encountered in their everyday experience.

In short, naturalists before Linnaeus and Fabricius found it devilishly hard to reconcile their new diachronic concept of the insect species with the increasing synchronic attention to the immense diversity of the insect world.

[3] F. S. Bodenheimer, *Materialien zur Geschichte der Entomologie bis Linné*, 2 vols. (Berlin: W. Junk, 1928–29), 1: 7–9, 312–498.

[4] Marcel Dicke, "Insects in Western Art," *American Entomologist* 46.4 (2000): 228–237.

[5] The term "insect world" seems to come into use in the late 18th century, e.g., in the title of a 1791 abridged translation of Buffon (Adam Dodd, personal communication).

As Réaumur put it in 1734, "The immensity of nature's work appears nowhere more clearly than in the innumerable multiplicity of species of little animals."[6] In the attempt, though, they identified patterns that their successors would draw upon to create a new science that they would eventually call, after a halting start, "entomology."[7] In this chapter, I trace the history of the attempt to understand insect species, diversity, and metamorphosis from the late Renaissance to the early Enlightenment. My account is necessarily selective. In particular, the role of visual representations in understanding insects is touched on only briefly.[8] Nor do I address in any depth the material basis of insect studies: collections, experiments, and dissection. Nonetheless, the related problems of diversity and metamorphosis provide a useful thread for understanding how early modern Europeans thought about one aspect of the order of nature.

2. Renaissance Classification

Renaissance naturalists were aware of insect metamorphosis. But their classifications downplayed or ignored it, emphasizing instead the synchronic morphology of individual specimens, and placing larval and adult forms in different parts of their classifications. This approach was based in naturalists' collecting practices, but it was justified as well by their views about the nature of insect generation, in particular, the belief in equivocal or spontaneous generation. Though many 16th-century scholars, both well known (like Conrad Gessner and Thomas Penny) and unknown (like Jacob Zwinger and Pascal Le Coq), were working on histories of insects, I focus my brief overview of Renaissance insectology on three important publications: Edward Wotton's *De differentiis*

[6] René-Antoine Ferchauld de Réaumur, *Mémoires pour servir à l'histoire des insectes*, 6 vols. (Paris: Imprimerie Royale, 1734–42), 1: 2.

[7] The word "entomologia" appears to have been coined by Johann Heinrich Alsted in 1630, but it came into widespread use only in the wake of Linnaeus's *Systema naturae*. Johann Heinrich Alsted, *Encyclopaedia septem tomis distincta* (Herbornae Nassoviorum: [s.n.], 1630), 1: 794, 796. The first book to use the word in its title was Giovanni Antonio Scopoli, *Entomologia carniolica exhibens insecta Carnioliae indigena et distributa in ordines, genera, species, varietates methodo Linnæana* (Vindobonae: Typis Ioannis Thomae Trattner, Caes. Reg. Aulae Typog. et Bibliop., 1763).

[8] On visual representation, see Thea Vignau-Wilberg, "In minimis maxima conspicua. Insektendarstellungen um 1600 und die Anfänge der Entomologie," in *Early Modern Zoology: The Construction of Animals in Science, Literature and the Visual Arts*, eds. Karl A. E. Enenkel and Paul J. Smith (Leiden, NL, and Boston: Brill, 2007), 1: 217–243; Janina Wellmann, "Die Metamorphose der Bilder: Die Verwandlung der Insekten und ihre Darstellung vom Ende des 17. bis zum Anfang des 19. Jahrhunderts," *N.T.M.* 16 (2008): 183–211; Domenico Bertoloni Meli, "The Representation of Insects in the Seventeenth Century: A Comparative Approach," *Annals of Science* 67.3 (2010): 405–429; and Janice Neri, *The Insect and the Image* (Minneapolis: University of Minnesota Press, 2011).

animalium (*On the differentiating characters of animals*, 1552) and the two great *summae* of Renaissance insect investigation and lore, Ulisse Aldrovandi's *De animalibus insectis libri septem* (*Seven books on insects*, 1602) and Thomas Moffett's *Insectorum sive minimorum animalium theatrum* (*Theater of Insects, i.e., of the smallest animals*, published in 1634 but largely complete by the 1590s).[9]

Edward Wotton (1492–1555) appears not to have published many, if any, original observations of animals; his *De differentiis animalium* was compiled from ancient sources.[10] As Karl Enenkel has emphasized, Wotton's goal was to produce a systematic compendium based on Aristotle's method of dividing animals.[11] Following Aristotle's classification, Wotton first divided animals into those with blood and without; the latter, in turn, were subdivided into insects, *mollia* (soft-bodied creatures), crustaceans, and hard-shelled creatures.[12] Again following Aristotle, Wotton defined "insects" as those creatures whose bodies were incised, either on top, on the bottom, or both, and divided into three parts: a head, a belly, and a connecting part, which might be multiplied in the case of centipedes and related creatures.[13]

Within the broad class of insects, however, Wotton did not draw finer distinctions. In keeping with folkbiological perceptions of similarity, he noted that some insects belonged together even if there was no common name for them: bees and other comb-making insects formed a "certain genus of insects that thus far lacks a name."[14] After discussing bees, wasps, hornets, and bumblebees, however, Wotton turned to a hodgepodge of forms: silkworms, ants, scorpions, spiders, centipedes and millipedes, cicadas, locusts, beetles, caterpillars,

[9] For Zwinger and Le Coq, see Basel University Library (Switzerland), MS. Fr. Gr. I. 12, nos. 123 and 274; and MS. Fr. Gr. I. 13, nos. 64, 75, and 82. I use the term "insectology," which was coined by Charles Bonnet in his 1745 *Traité de l'insectologie* but never caught on, to defamiliarize the study of insects. As noted above, "entomology" comes into widespread use only in the wake of Linnaeus.

[10] On Wotton, see Willy Ley, *Dawn of Zoology* (Englewood Cliffs, NJ: Prentice-Hall, 1968); Änne Bäumer, "Das erste zoologische Kompendium in der Zeit der Renaissance: Edward Wottons Schrift 'Über die Differenzen der Tiere,'" *Berichte zur Wissenschaftsgeschichte* 13 (1990): 13–29; Änne Bäumer, "Edward Wotton: Aristotelische Zoologie aristotelischer als bei Aristoteles," in *Vorträge des ersten Symposions des Bamberger Arbeitskreises "Antike Naturwissenschaft und ihre Rezeption,"* eds. Klaus Döring and Georg Wöhrle (Wiesbaden, DE: Harrassowitz, 1990), 187–211; A. F. Pollard and Patrick Wallis, "Wotton, Edward (1492–1555)," *Oxford Dictionary of National Biography*, 2004, http://www.oxforddnb.com/view/article/29999; Karl A. E. Enenkel, "Zur Konstituierung der Zoologie als Wissenschaft in der Frühen Neuzeit: Diskursanalyse zweier Großprojekte (Wotton, Gesner)," in *Early Modern Zoology*, eds. Enenkel and Smith, 1: 15–74.

[11] Enenkel, "Zur Konstituierung der Zoologie," 28–31.

[12] Edward Wotton, *De differentiis animalium libri decem* (Lutetiae Parisiorum: apud Vasconsanum, 1552), fol. 173v–4r.

[13] Wotton, *De differentiis animalium*, fol. 175r.

[14] Wotton, *De differentiis animalium*, fol. 177v: "Insectorum genus quoddam est nomine adhuc vacans quo uno comprehendatur."

and then "bestiolae" that were born in wood, plants, inanimate material, and inside other creatures. He ended with those that lived off of humors (fleas, lice, ticks, etc.) and, oddly, roaches.[15] There is no implicit subdivision of insects within this discussion.

Wotton did recognize that insects metamorphose. He devoted a chapter to insects' generation and death, while another specifically addressed "caterpillars and the creatures that are born from them, such as the butterflies, and certain other winged creatures that are born from worms."[16] Following his classical authorities, however, Wotton saw that kind of generation as only one way in which insects arose. Some, like spiders and grasshoppers, produced offspring similar to themselves, while others, such as flies and butterflies, generated offspring that were different from themselves. Others, meanwhile, were generated spontaneously from decaying matter or even from dew, vinegar, or pure snow.[17] This commitment to spontaneous or equivocal generation would shape the approach taken to insects by Wotton's late humanist successors Moffett and Aldrovandi, who produced insect classifications that drew a sharp distinction among forms that later naturalists would consider to be successive stages in the development of the same species.[18]

Unlike Wotton, who wove his book entirely out of excerpts from classical sources, Ulisse Aldrovandi approached insects not only as textual objects, but also as individual specimens that he observed, collected, had drawn, and sometimes killed and brought back to form part of his "diverse collection (*variam supellectilem*)."[19] Despite his attention to insects' generation, Aldrovandi's approach to their classification was essentially synchronic, based on the place where each insect was found and the form of its body. The synoptic table that opened his *De animalibus insectis* first divided insects into aquatic and terrestrial, then proceeded further according to whether they lacked feet or had them, and among the latter, whether they had wings or not.[20] It is tempting to see this approach as a way to organize drawings and specimens. One of its consequences was that butterflies and moths belonged to the terrestrial insects with feet and wings, while the caterpillars from which they were born belonged to distinct groups, the terrestrial insects with 12 or 14 feet.

[15] Wotton, *De differentiis animalium*, fol. 177v–97r.

[16] Wotton, *De differentiis animalium*, fol. 176r, 193r: "De erucis, & quae ex iis gignuntur, ut papilionum generibus, & de pennatis quibusdam aliis quae ex vermibus gignuntur."

[17] Wotton, *De differentiis animalium*, fol. 176r–6v.

[18] On this point, see Matthew Cobb, *Generation: The Seventeenth-Century Scientists Who Unraveled the Secrets of Sex, Life, and Growth* (New York: Bloomsbury USA, 2006).

[19] Ulisse Aldrovandi, *De animalibus insectis libri septem* (Bonon: apud Ioan. Bapt. Bellagambam, 1602), sig. †3r.

[20] Aldrovandi, *De animalibus insectis*, sig. †4r.

In practice, Aldrovandi deviated from this scheme when composing his book; recognizing that caterpillars gave rise to butterflies, he included both in Book 2 of *De animalibus insectis,* which was devoted to certain "four-winged insects without wing-cases." He occasionally noted which butterflies developed from certain caterpillars, including cross references: For example, "When I had nourished [the first caterpillar of the first table] for some time in my house, it did not weave a web or sack, but rather formed a chrysalis which gave birth to a darkish yellow butterfly, namely the third in the first table." And like Wotton, he accepted Aristotle's principle that butterflies tend to have the same color as their caterpillars.[21] There is even one sheet in Aldrovandi's papers in Bologna that suggests he started—but abandoned—a systematic study of the relationship between species of caterpillars, chrysalises, and adult butterflies and moths.[22]

Aldrovandi's commitment to spontaneous generation encouraged his synchronic, morphological approach to insects. He cautiously noted that he had observed butterflies laying eggs that, in turn, became caterpillars, though he left it to his reader's judgment as to whether this happened regularly. Moreover, this did not happen all the time: Many caterpillars died without giving rise to anything, while others produced a chrysalis that gave rise not to a winged insect but to several worms.[23] The view held by proponents of equivocal generation that insects were imperfect animals arising out of putrefaction applied even to the transformation from caterpillar to butterfly or from grub to beetle; the wingless creature, dying, gave rise to an "egg" (the pupa) that produced a wholly new creature, in the process that William Harvey, who would reject it, called "metamorphosis."[24] Francis Bacon reflected this belief in one of the aphorisms of his *Sylva sylvarum*: "We see that in living creatures that come of putrefaction, there is much transmutation of one into another; as caterpillars turn into flies, etc. And it should seem probable that whatsoever creature, having life, is generated without seed, that creature will change out of one species into another."[25] For Bacon, this kind of transmutation was one of nature's "mighty works (*magnalia*)" that called

[21] Aldrovandi, *De animalibus insectis,* 267: "Cum eam aliquo tempore domi aluissem, telam non texuit, aut folliculum, sed in Chrysalidem immutata Papilionem peperit atro luteum, eum quem in prima Papilionum tabula tertium, exhibuimus."

[22] Janice Neri and I hit on this idea independently. Janice Neri, "Fantastic Observations: Images of Insects in Early Modern Europe" (Ph.D. diss., University of California, Davis, 2003), 53; Neri, *Insect and the Image,* 34; and Brian W. Ogilvie, "Nature's Bible: Insects in Seventeenth-Century European Art and Science," *Tidsskrift for kulturforskning* 7.3 (2008): 9.

[23] Aldrovandi, *De animalibus insectis,* 253–256.

[24] See Cobb, *Generation,* 132–133.

[25] Francis Bacon, "Sylva Sylvarum; or, A Natural History," in *The Works of Francis Bacon,* eds. James Spedding, Robert Leslie Ellis, and Douglas Denon Heath, vol. 4 (Boston: Brown and Taggard; Taggard and Thompson, 1862), 400.

into question the "vulgar philosophy" whose proponents held the transmutation of species to be impossible.

In the *Insectorum theatrum*, published after Aldrovandi's work but probably completed before it, Moffett adopted a classification that adumbrates (or copies) Aldrovandi's. Except for scorpions, which Moffett lumped with ants, the first book is devoted to winged insects, beginning with bees, wasps, and their ilk, then discussing flies, butterflies and moths, locusts, cicadas, crickets, beetles, crane flies, and more. The second book, in turn, contains the wingless insects.[26] Its first 28 chapters address terrestrial wingless insects with feet, beginning with caterpillars before turning to millipedes, centipedes, woodlice, spiders, ants (without wings), and hexapods (beetle larvae); the remaining 14 are devoted to legless and aquatic insects. As in Aldrovandi's classification, larval and adult forms are separated: Caterpillars are found in book 2, chapters 1–5, while butterflies are in book 1, chapter 14.

Moffett (or his sources), like Aldrovandi, noted that certain caterpillars regularly transformed into butterflies; some of his images of caterpillars included the associated chrysalis and adult.[27] But again like Aldrovandi, he held that this generation was not entirely regular. Some caterpillars died and putrefied without transforming, while others produced black eggs that became flies.[28] The chrysalis, in turn, seemed to be quite anomalous. Aristotle had excluded it from the realm of animals, considering it to be the caterpillar's egg. Moffett thought it was not an egg, for it was not laid by another creature; like a hibernating dormouse it was alive, even though it neither moved nor ate. However, it was a stage in a transmutation between different creatures, from caterpillar to chrysalis to butterfly.[29] For Moffett, as for Aldrovandi, the diachronic sequence of life stages was less important than the creature's form and dwelling place.

Renaissance naturalists thus recognized the fact of insect metamorphosis but subordinated it to other concerns. They were justified in doing so by their practice of describing and depicting creatures synchronically, but also by their commitment to equivocal generation: "Metamorphosis," in William Harvey's sense, was the sudden crystallization of form from formlessness, something

[26] Thomas Moffett, *Insectorum sive minimorum animalium theatrum: Olim ab Edoardo Wottono, Conrado Gesnero, Thomaque Pennio inchoatum: tandem Tho. Moufeti Londinatis opera sumptibusq; maximis concinnabum, auctum, perfectum* (Londini: ex officina typographica Thom. Cotes, et venales extant apud Guiliel. Hope, 1634), 178. On Moffett, see Charles E. Raven, *English naturalists from Neckam to Ray: A study of the making of the modern world* (Cambridge: Cambridge University Press, 1947), 172-191; Victor Houliston, "Moffet, Thomas (1553–1604)," *Oxford Dictionary of National Biography*, 2004, http://www.oxforddnb.com/view/article/18877; and Neri, *Insect and the Image*, 45-73.

[27] E.g., Moffett, *Insectorum theatrum*, 106, 184, 186.

[28] Moffett, *Insectorum theatrum*, 192.

[29] Moffett, *Insectorum theatrum*, 318–319.

that happened only in imperfect creatures. The new approaches to insect classification that developed in the late 17 century would depend on a fundamental rethinking of that process, including the rejection of spontaneous generation. But it would also depend on the painstaking empirical work of the men—and at least one woman—who demonstrated the regular change of each wingless insect into its winged counterpart.

3. The Study of Metamorphosis: Willughby, Goedaert, Lister, and Merian

The first significant classification of insects after Aldrovandi's and Moffett's was produced by the young English virtuoso Francis Willughby.[30] It appeared not in a work of natural history but in John Wilkins's *Essay towards a Real Character and a Philosophical Language*, an ambitious if not utopian attempt to produce a universal language.[31] Wilkins required a taxonomy of all natural kinds, and he engaged Willughby to provide a classification of animals, including insects, in which Willughby had a particular interest.[32]

Willughby's taxonomy reflected a new interest in problems of generation.[33] In the wake of William Harvey's path-breaking *On the Generation of Animals* (1651), physicians and naturalists took a new interest in how animals reproduced. Willughby had been studying insects' generation, in particular the metamorphosis of caterpillars, and he used generation to create the basis of his classification of insects. He first divided them broadly into those that, on the

[30] On Willughby, see G. S. Boulger and Michael Hunter, "Willughby, Francis (1635–1672)," *Oxford Dictionary of National Biography*, 2004, http://www.oxforddnb.com/view/article/29614; and the biography in Francis Willughby et al., *Francis Willughby's book of games: A seventeenth-century treatise on sports, games, and pastimes* (Aldershot, UK, and Burlington, VT: Ashgate, 2003).

[31] John Wilkins, *An Essay Towards a Real Character and a Philosophical Language* (London: Printed for Sa: Gellibrand, and for John Martyn, printer to the Royal Society, 1668); on Wilkins's project, see Mary M. Slaughter, *Universal Languages and Scientific Taxonomy in the Seventeenth Century* (Cambridge, UK: Cambridge University Press, 1982); Sidonie Clauss, "John Wilkins' Essay toward a Real Character: Its Place in the Seventeenth-Century Episteme," *Journal of the History of Ideas* 43.4 (1982): 531–553; William T. Stearn, "The Wilkins Lecture 1985: John Wilkins, John Ray and Carl Linnaeus," *Notes and Records of the Royal Society of London* 40.2 (May 1986): 101–123; R. Lewis, "The Publication of John Wilkins's 'Essay' (1668): Some Contextual Considerations," *Notes and Records of the Royal Society of London* 56.2 (May 2002): 133–146.

[32] Ray attested to Willughby's interest in insects in the preface to his publication of *The Ornithology of Francis Willughby* (1676 Latin, 1678 English) and elsewhere; see Brian W. Ogilvie, "Attending to Insects: Francis Willughby and John Ray," *Notes and Records of the Royal Society* 66.4 (2012): 357–372.

[33] See Jacques Roger, *Les sciences de la vie dans la pensée française du XVIIIe siècle: La génération des animaux de Descartes à l'Encyclopédie* (Paris: A. Colin, 1963); and Cobb, *Generation.*

one hand, had a generation "analogous to that of other Animals; which breed young like themselves, growing from a lesser to a greater magnitude, not being transmutable into any other *Insects*," and those that, on the other, have an "anomalous" generation. These latter, in turn, are divided into two kinds: those that "are designed to a further tra[n]smutation," such as maggots, caterpillars, and the like, and those that have already gone through such a transmutation, from egg to larva to pupa to adult.[34]

Despite his attention to generation, Willughby retained the Renaissance emphasis on synchronic form. Each insect that metamorphosed was included twice, once in its larval form and once again in its adult form. It is possible that this duplication was Wilkins's idea: The year after the *Real Character* was published, John Ray complained to Martin Lister that the tables of plants he had done for Wilkins were accommodated to the author's system, not to nature.[35] But it seems likely that Willughby, in keeping with his predecessors, could not bring himself to class in the same group such radically different forms as a 14-legged caterpillar and a six-legged, winged butterfly; in the 1670s, Ray too thought that Aldrovandi's classification, which also separated larval and adult forms, was "no ill one."[36]

As Jan Swammerdam, Martin Lister, René-Antoine Ferchault de Réaumur, and others recognized, the investigator who did more than anyone else to convince 17th-century insectologists that insects must be considered diachronically was Johannes Goedaert (1617–68).[37] A professional painter, Goedaert spent his entire life in the Dutch town of Middelburg. In the 1630s, he began to collect caterpillars and other insect larvae, raise them in glass containers, and study their "strange transformations"; between 1660 and his death, he described those changes in three volumes titled *Metamorphosis naturalis, ofte Historische*

[34] Wilkins, *Essay Towards a Real Character*, 122.

[35] John Ray, *The Correspondence of John Ray: Consisting of Selections from the Philosophical Letters Published by Dr. Derham, and Original Letters of John Ray, in the Collection of the British Museum* (London: Printed for the Ray Society, 1848), 30.

[36] Ray to John Aubrey, some time in the second half of 1678, in John Ray, *Further Correspondence of John Ray* (London: Printed for the Ray Society, 1928), 165.

[37] On Goedaert, see G. Kruseman, "The Editions of Goedaert's Metamorphosis naturalis," *Entomologische Berichten* 16 (1956): 46–48; L. J. Bol, "Een Middelburgse Brueghel-groep. IX. Johannes Goedaert, schilder-entomoloog," *Oud Holland* 74 (1959): 1–19; Peter W. Van der Pas, "Goedaert, Johannes (b. Middelburg, Netherlands, ca. 19 March 1617; d. Middelburg, February 1668)," in *Dictionary of Scientific Biography*, ed. Charles Coulton Gillespie (New York: Charles Scribner's Sons, 1970–80), 5: 439–440; L. J. Bol, "Johannes Goedaert, schilder-entomoloog [3 parts]," *Tableau* 7 (1984-85): no. 2, 65–70; no. 3, 64–69; no. 4, 48–54; Anita Albus, *The Art of Arts: Rediscovering Painting* (New York: Alfred A. Knopf, 2000); Neri, "Fantastic Observations"; Eric Jorink, "Between emblematics and the 'argument from design': The representation of insects in the Dutch republic," in *Early Modern Zoology*, eds. Enenkel and Smith, 1: 147–175; and my discussion in Ogilvie, "Nature's Bible," from which the following is adapted.

beschryvinghe vanden oirspronk, aerd, eggenschappen ende vreemde veranderinghen der wormen, rupsen, maeden, vliegen, witjens, byen, motten ende dierghelijcke dierkens meer (*Natural Metamorphosis, or Historical description of the origin, nature, characters and strange transformations of worms, caterpillars, maggots, flies, whites, bees, moths, and other such little creatures*).[38] Though the work was written in Dutch, Goedaert's publisher also issued a Latin translation, ensuring that it was available across Europe; Martin Lister published an English translation in 1682, and a French edition appeared in 1700.

Goedaert did not classify insects. He simply observed their behavior and metamorphosis, drawing them as larvae, pupae, and adults. His achievement was to emphasize the diachronic aspect of insects' lives. Many of them did not simply grow from smaller to larger forms; rather, they underwent a striking transformation. And yet they seemed to be the same creature. Some of the transformations were more striking than others: In one case, Goedaert noted that a certain caterpillar sometimes gave birth to a butterfly, but at other times to several worms that, in turn, transformed themselves into flies. "These things I have had the experience of, and Observed them, not without admiration," Goedaert wrote, "because it seems besides, if not against the usuall course of Nature, that from one and the same Species of Animals, an Offspring of different species shou'd be gendred."[39]

Goedaert's painstaking researches earned him a reputation as the person to go to if one wanted to know what kind of caterpillar had produced a butterfly: One of his few illustrations with only an adult form depicted a butterfly from the Jardin du Roy in Paris; the Dutch ambassador to Versailles had sent it to Goedaert for an identification.[40] Yet the publication of his work in Latin, and later in French, suggested that Goedaert's researches raised broader questions about "the usuall course of Nature." Indeed, Leibniz thought that Goedaert's insects belonged in a public curiosity cabinet.[41] As Martin Lister realized when he encountered Goedaert, the Dutch artist's approach underscored that the study of insects must be irreducibly diachronic. It must take into account the

[38] Johannes Goedaert, *Metamorphosis naturalis, ofte Historische beschryvinghe vanden oirspronk, aerd, eggenschappen ende vreemde veranderinghen der wormen, rupsen, maeden, vliegen, witjens, byen, motten ende dierghelijcke dierkens meer*, 3 vols. (Middelburgh: Jaques Fierens, Boeck-verkooper, inde Globe, 1660–69); on the work's publication dates, see Ella Reitsma, *Maria Sibylla Merian and daughters: Women of art and science* (Amsterdam: The Rembrandt House Museum; Los Angeles: J. Paul Getty Museum; Zwolle: Waanders Publishers, 2008), 68n22.

[39] Johannes Goedaert, *Of insects: Done into English and Methodized with the Addition of Notes* (York: Printed by John White for M. L., 1682), 10–11.

[40] Goedaert, *Of insects*, 37–38.

[41] Philip P. Wiener, "Leibniz's Project of a Public Exhibition of Scientific Inventions," *Journal of the History of Ideas* 1.2 (1940): 237.

entire life cycle of the animal, now considered as one species under several distinct forms. But it must also be systematized.

On the positive side of the ledger, Lister thought that Goedaert was a diligent observer who noted and drew only what he himself had seen, "which is more, than any man ever did before him; So that we need not admire, that so long, and pertinatious an Industry produced so few Historys: For he Designed not all, that came to hand, but such only, as it was his good fortune to Feed, and bring up to Change. And yet in these few Historys, you will have something of all the severall Genus's of Insects, that are in Nature."[42] Though Goedaert's work was not comprehensive, it provided, perhaps unintentionally, a broad overview of insect metamorphosis.

But Goedaert needed to be criticized and systematized. Lister rearranged the chapters of Goedaert's three volumes systematically, including cross-references to the original Latin edition. He sharply criticized some of Goedaert's more credulous stories. Most importantly, though, Lister's study of ichneumon wasps, part of his research on spiders, had revealed to him the true nature of Goedaert's changes that were contrary to nature's usual course: They resulted from parasitism by wasps that somehow introduced their eggs into the caterpillar's body.[43]

For example, Goedaert described one caterpillar that changed into a chrysalis, "and when it had been some dayes in that condition, Three Wormes broke out through the skin of it." These worms in turn became "eggs," each of which produced a "Fly" that lived a few days. Lister commented acerbically, "These were Flesh-flyes, and consequently the Chrysalis was rotten, and this History not finished; because we have not the Butterflyes, which in reason we ought to expect."[44] As a result of such errors, Lister thought, Goedaert did not succeed in truly capturing the essence of an insect.

Still, the lesson that metamorphosis was not a change between species but a change within one was adopted by Lister. When he reorganized Goedaert's material, he classified them by broad groups. The first two of them explicitly drew on characters from two phases of the insect's life cycle: "Butterflyes, sitting with erect Wings: all whose Chrysoli's are Angular: of which there are VIII. species," followed by "Butterflyes sitting with flat Wings, whose Caterpillers want the middle Leggs, and from there manner of going in Loops, are by some called Geometrae." From a modern standpoint, Lister's terms might seem odd, such as his group "Of Butterflys with Transparent wings, known by the name of Draggon Flyes, in English." And his other groups—bees, beetles, grasshoppers, flies, millipedes, and spiders—correspond to widely accepted folk divisions. But

[42] Lister in Goedaert, *Of Insects*, sig. A3r–v.

[43] Anna Marie Roos, *Web of Nature: Martin Lister (1639–1712), the First Arachnologist* (Leiden and Boston: Brill, 2011), 93–94, 123–124.

[44] Goedaert, *Of Insects*, 22.

the idea that species and larger taxa of insects would need to consider life cycle, not merely morphology, was established and would continue to gain currency.

Martin Lister's response to Goedaert was to reorganize his descriptions and illustrations and critique his naïveté. Maria Sibylla Merian took another tack: She far exceeded her predecessor in making acute observations and producing precise, elegant depictions of metamorphosis. Merian was born in Frankfurt, where her stepfather, the artist Jacob Marrel, trained her in his craft.[45] As she told the story, she began looking at insects so that she could adorn her pictures with them, but on discovering their transformations, she decided to study the matter more carefully.[46] She began this work as at the age of 13 and continued it for over five decades, first in Frankfurt, then Nuremberg, the Labadists' religious colony in Wieuwerd (Frisia), Amsterdam, and Suriname, where she spent nearly two years, from the summer of 1699 to the summer of 1701.

Merian kept careful notes and made exquisite drawings of the insect larvae that she collected and raised in order to observe their metamorphoses.[47] Her studies drew attention from the learned circle in Nuremberg in which she and her husband, the painter Johann Andreas Graff, participated after their move to that city. In his 1675 *Teutsche Akademie*, the artist and art historian Joachim von Sandrart described her interest in insects and skill in depicting them.[48] Another member of the circle, the theologian and poet Christoph Arnold, may have introduced her to the growing scholarly literature on insects. He was probably one of the "learned and distinguished people" who encouraged her to publish a book of her observations: *Der Raupen wunderbare Verwandelung und sonderbare Blumen-nahrung* (*The wondrous transformation and strange flower-nourishment of*

[45] There is a large and growing literature on Merian. I have found the following works most helpful: Natalie Zemon Davis, *Women on the Margins: Three Seventeenth-Century Lives* (Cambridge, MA: Harvard University Press, 1995); Kurt Wettengl, ed., *Maria Sibylla Merian: Künstlerin und Naturforscherin, 1647–1717* (Ostfildern-Ruit, DE: Verlag Gerd Hatje, 1997); Helmut Kaiser, *Maria Sibylla Merian: Eine Biographie* (München and Zürich: Piper, 1999); Kim Todd, *Chrysalis: Maria Sibylla Merian and the Secrets of Metamorphosis* (Orlando, FL: Harcourt, 2007); Reitsma, *Merian and Daughters*; Anne-Charlott Trepp, *Von der Glückseligkeit alles zu wissen: Die Erforschung der Natur als religiöse Praxis in der Frühen Neuzeit* (Frankfurt am Main: Campus, 2009); and Neri, *Insect and the Image.*

[46] Maria Sibylla Merian and Johanna Helena Herolt, *Erucarum ortus, alimentum et paradoxa metamorphosis*, 3 pts. (Amstelaedami: Apud Joannem Oosterwyk, 1718), 1: sig. *3r.

[47] Her studies and notes in St. Petersburg, usually referred to by Merian scholars as the Study Book, have been published in facsimile: Maria Sibylla Merian, *Butterflies, Beetles and Other Insects: The Leningrad Book of Notes and Studies* (McGraw-Hill International Book Co., 1976). Other important collections of Merian's original studies are in the Royal Collection in Windsor and in the British Library.

[48] Reitsma, *Merian and Daughters*, 23–24.

caterpillars), which appeared in 1679.[49] This was the first of Merian's three publications on insect metamorphosis.[50]

Merian's decision to publish her observations may have been indirectly inspired by Goedaert. She does not mention him in the preface to her first caterpillar book, but Arnold's dedicatory poem compares Merian favorably to the Dutch artist—as well as to Moffet, Swammerdam, Harvey, Redi, and others.[51] Many of her studies of insect metamorphosis follow the same pattern as Goedaert's engravings.[52] But Merian was a more careful observer than Goedaert. She discovered insect parasitism, noting in her first caterpillar book that some pupae produced "flies," not butterflies or moths, and eventually observing a wasp depositing her eggs on a caterpillar as it was about to pupate.[53] But her care also extended to other aspects of insects' life cycles.

Insect larvae molt several times as they grow, shedding their old skin and forming a new one; these successive stages are called instars. In many cases, the new instar appears to be a bigger version of the old, though the coloration can change as the new skin dries and hardens. In other species, successive instars can differ radically in morphology and coloration. Merian described and depicted both kinds of change, in caterpillars and beetle larvae. Where Goedaert had shown only one form for each of his larvae, Merian included many if the instars differed significantly.[54] She also noted a few cases in which male and female insects had different forms, though in most cases, sexual differences required using a hand lens to identify.[55]

[49] Maria Sibylla Merian, *Der Raupen wunderbare Verwandelung und sonderbare Blumen-nahrung*, 2 vols. (Nürnberg; Frankfurt; Leipzig: gedruckt bey Andreas Knortzen, 1679–83), 1: sig.):(iiir.

[50] I am counting the 1679 caterpillar book; the second volume, which was published in 1683, and the 1705 book on Surinamese insects, published simultaneously in Dutch and Latin versions. The two German volumes were later published in an abridged Dutch translation, with a third volume containing additional observations and illustrations of Dutch insects done by Merian and her elder daughter Johanna Helena; a Latin translation appeared shortly after Merian's death: Maria Sibylla Merian and Johanna Helena Herolt, *Der Rupsen begin, voedzel en wonderbaare verandering*, 3 vols. (Amsterdam: Gedrukt voor den Auteur [parts 1–2]; Gedrukt voor de Uytgeefster [part 3], 1713–17); Merian and Herolt, *Erucarum ortus*.

[51] Merian, *Der Raupen wunderbare Verwandelung*, 1: t.p. recto. The "Bustamantin" whom Arnold mentioned is Juan Bustamante de la Cámara, author of *De animantibus sacrae scripturae* (1595–1620); cf. Reitsma, *Merian and Daughters*, 72.

[52] Neri, *Insect and the Image*, 157.

[53] Merian, *Der Raupen wunderbare Verwandelung*, 1: 12; Merian and Herolt, *Erucarum ortus*, 1: sig. *3v. J. C. van Lenteren and H. C. J. Godfray, "European science in the Enlightenment and the discovery of the insect parasitoid life cycle in The Netherlands and Great Britain," *Biological Control* 32.1 (2005): 12–24, suggest that the second observation might have been made between 1686 and 1691, the years Merian spent at the Labadist colony in Wieuwerd, but they have no direct evidence.

[54] E.g., Merian, *Der Raupen wunderbare Verwandelung*, 1: 52, 55–56; Maria Sibylla Merian, *Metamorphosis insectorum surinamensium, in qua erucae ac vermes surinamenses cum omnibus suis transformationibus ad vivum delineantur et describuntur* (Amstelodami: Sumtibus Auctoris, 1705), 11.

[55] Davis, *Women on the Margins*, 148.

The most striking aspect of Merian's insect publications, however, is their form. While many of the metamorphoses depicted in Merian's notes show larvae, pupae, and imagos against a blank background, her publications—the 1679 caterpillar book, its 1683 successor, and her spectacular 1705 *Metamorphosis insectorum surinamensium* (*Metamorphosis of the insects of Suriname*)—depict them on or near plants, often, though not always, a food plant of the larva. The title of her 1679 caterpillar book underscores her interest in insects' foodstuffs, representing what has been called an "ecological" perspective.[56] Merian's compositions also drew on her experience producing floral patterns for embroidery.[57] By combining elegant floral compositions with accurately portrayed insects at each stage of their metamorphoses, Merian produced works that were prized by both art collectors and naturalists.

Unlike Lister, Merian had no interest in classifying or even naming insects. Linnaeus would list her, along with Goedaert and other artists, as an expert on insect metamorphosis.[58] But she preferred to leave speculation and system-building to others. In the preface to her book on the insects of Suriname, she explained, "I could have written more, but the present-day world is very delicate and the learned have differing feelings, so I prefer to simply stick with my observations, and thereby deliver material that everyone can reflect on according to his understanding and opinions."[59] There was too much of that already, she continued; it could be found in Moffett, Goedaert, Blankaart, and in one other writer whom Linnaeus would also praise: Swammerdam.

4. Classification and Metamorphosis: Swammerdam, Ray, Vallisnieri, and Réaumur

Like Merian, Jan Swammerdam was fascinated by metamorphosis; unlike his German contemporary, whose work he never encountered, Swammerdam attempted to find a deeper pattern lying beneath the multiplicity of transformations.[60] Swammerdam, a talented, mercurial anatomist, produced incredibly

[56] Davis, *Women on the Margins*, 151.

[57] Neri, *Insect and the Image*, 142–154.

[58] Carolus Linnaeus, *Systema naturae per regna tria naturae, secundum classes, ordines, genera, species, cum charateribus, differentiis, synonymis, locis*, 2 vols. (Holmiae: Impensis L. Salvii, 1758–59), 1: 341.

[59] Maria Sibylla Merian, *Metamorphosis insectorum surinamensium, ofte Verandering der surinaamsche insecten* (Amsterdam: Voor den Auteur, 1705), unpaginated letter to the reader.

[60] On Swammerdam, see Matthew Cobb, "Reading and Writing the Book of Nature: Jan Swammerdam, 1638–1680," *Endeavour* 24.3 (2000): 122–128; Matthew Cobb, "Malpighi, Swammerdam and the Colourful Silkworm: Replication and Visual Representation in Early Modern Science," *Annals of Science* 59 (2002): 111–147; F. J. Cole, "The Wilkins Lecture: History of Micro-Dissection," *Proceedings of the Royal Society of London, Series B, Biological Sciences* 138.891

detailed anatomical and behavioral studies of particular insects—the mayfly, the bee, the louse—while insisting on the necessity of a diachronic view of insect life. Though much of his work remained unpublished until it was rediscovered by Hermann Boerhaave, who gave it the title *Bybel der natuure* (*Bible of Nature*), had it translated into Latin, and published it in 1737–38, Swammerdam's classification was laid out in his 1669 *Historia insectorum generalis* (*General History of Insects*, which was in Dutch, its title notwithstanding) and its French (1682) and Latin (1685) translations.[61]

Swammerdam was deeply critical of his predecessors, from Aristotle and William Harvey to Moffett and Goedaert, for their false beliefs about the nature of insects' metamorphoses: in particular the widespread idea, which I have already mentioned, that insect larvae or "worms" arose from putrefaction as imperfect animals, died in forming a chrysalis or perfect egg, which in turn gave birth to the perfected animal, that is, the adult insect. Although he credited some of his predecessors, such as Goedaert and Harvey, with being careful observers, he believed that their care had failed them in this case, for three reasons. First, as Francesco Redi had shown, insects did not generate spontaneously, but rather grew from eggs. Second, those theories failed to account for insects that hatched from eggs in a form that closely resembled the adult. And third, Swammerdam's own observations showed that even insects that changed most radically—endopterygotes, or holometabolous insects, we would say—developed slowly and regularly from larva to imago. The caterpillar pupated when the organs that would characterize the adult insect had reached the proper point in development, and the pupa or chrysalis clearly revealed those organs.

Underlying Swammerdam's specific critiques was a general methodological principle. Given the weakness of human reason, observation is the only true way to knowledge of nature:

> Now all of God's works are based on the same rules, and we find the true causes and origins of them to be completely impenetrable: hence in truth we know only the outline of some shadows of his marvels. Therefore we hold that the only knowledge we can have is of clear appearances or effects, which in turn are often causes of other effects. In

(June 28, 1951): 159–187; Marian Fournier, "The Book of Nature: Jan Swammerdam's Microscopical Investigations," *Tractrix* 2 (1990): 1–24; Jorink, "Between Emblematics and Design"; Edward G. Ruestow, *The Microscope in the Dutch Republic: The Shaping of Discovery* (Cambridge, New York, and Melbourne: Cambridge University Press, 1996); and A. Schierbeek, *Jan Swammerdam, 12 February 1637–17 February 1680: His Life and Works* (Amsterdam: Swets & Zeitlinger, 1967).

[61] The catalogue of the Bibliothèque Nationale de France indicates a 1682 Latin edition in the Bibliothèque de l'Arsénal. I have not been able to consult it, but I suspect that it is a bibliographic ghost.

> order to know these, and to exhaust their positions, rules, conclusions, and reasons, we must employ all our diligence and effort.[62]

Observation was thus the only key to true philosophy; to prefer reason was a mark of being "puffed up with pride."[63]

But when Swammerdam looked at insect transformations, he saw the same patterns emerging again and again. He proposed "four orders which comprehend the whole class of insects, so that we cannot see one, which may not be referred to one or other of them, especially if we can see its change."[64] His first order comprises insects that hatch from eggs in a form that closely resembles the adult in all but size. In the second order, he places insects that become nymphs when the wings are close to perfection. The third involves those that as larvae have either no legs or six or more legs, and in which the adult's limbs grow under the skin until the larva is ready to molt and become a "Nymph, or Chrysalis." The fourth, finally, resembles the third except that the nymph forms underneath the larva's skin without a final molting.[65]

After describing the general characteristics of each order, Swammerdam listed the kinds of insects that he classifies in each one, and then provides a detailed description of a typical species, emphasizing its anatomy, followed by descriptions of other species. From our modern perspective, Swammerdam's focus on the kinds of metamorphosis makes strange bedfellows. In the first order, for instance, he located spiders, mites, lice, fleas, woodlice, worms, scorpions, leeches, centipedes, and snails. He was proud of having discovered how snails copulate, and he gives a thorough anatomical and behavioral explanation. The descriptions of animals in this class also include mussels.

The second order includes dragonflies. Of these he noted that "I preserve seventeen species of this genus."[66] It also includes grasshoppers, locusts, crickets, mole crickets, roaches, flying land and water bugs, and a few other "flies," including the mayfly. The dragonfly exemplifies the order's typical changes. The third order comprises the butterflies, night and day. The fourth, meanwhile, is an odd

[62] Jan Swammerdam, *Historia insectorum generalis, ofte Algemeene Verhandeling van de Bloedeloose Dierkens* (Utrecht: By Meinardus van Drevnen, Ordinaris Drucker van d'Academie, 1669), Naa-reeden, 5; Jan Swammerdam, *Histoire generale des insects: Ou l'on expose clairement la maniere lente & presqu'insensible de l'accroissement de leurs membres, & ou l'on decouvvre evidemment l'Erreur ou l'on tombe d'ordinaire au sujet de leur prétendué transformation* (Utrecht: Chez Jean Ribbius, 1685), 163.

[63] Swammerdam, *Historia insectorum generalis*, Naa-reeden, 10; Swammerdam, *Histoire generale des insectes*, 169.

[64] Jan Swammerdam, *The Book of Nature; or, The History of Insects: Reduced to Distinct Classes, Confirmed By Particular Instances, Displayed in the Anatomical Analysis of Many Species*, 2 vols. (London: Printed for C. G. Seyffert, 1758), 1: 17.

[65] Swammerdam, *Book of nature*, 1: 17–18.

[66] Swammerdam, *Book of nature*, 1: 93.

group. Swammerdam considered that it included the eggs of all insects, which he treats as a first nymph. It also includes insects that pupate inside other insects, like parasitic wasps; inside fruit, like the tiny wasps that fertilize figs; and indeed inside anything else.[67]

Swammerdam's classification shows how, despite his critique of received opinions, he retained the notions of imperfection and perfection. Except for the first, each of his four classes is characterized by an insect that is "imperfect" on emerging from the egg.[68] What distinguishes the second, third, and fourth classes is how that imperfection is rectified until a "perfect"—that is, adult—insect is produced. Nonetheless, Swammerdam was clear about their temporal relation: "It appears, from the nicest experiments, that all insects proceed from an egg, that is laid by an insect of the same species."[69] Swammerdam's original Dutch reads "uyt een Ey van diergelyk," translated in the 1737–8 *Bybel der Natuure* as "ex ovo, quod ab eiusdem speciei Insecto editum est."[70] Imperfection is now a stage in an insect's life, not an absolute quality of a lower being.

Swammerdam was much more interested in his broad classes than in individual species. After enumerating the different species of butterflies that his predecessors had described or depicted, he objected, "It would have been much more useful, to have exactly, and according to nature, described only one transformation of any Nymph, for an example to be applied to all the rest, than to have delineated the changes of all these Caterpillars, with their various colours, and their Nymphs: for, according to the other practice, the things which were most useful and necessary, remain untouched and neglected."[71] His predecessors, seduced by the diversity of insects, neglected their fundamental aspects.

Similarly, writing of moths (*Tinea* or *Mot-wurmen*), Swammerdam noted, "I should never finish, if I attempted to describe and delineate all the species of Moths, since there is a great number and variety of them. I should think, that what hath been hitherto said is sufficient, so that there is no need of more; for it is easy to know a Lion by his paw." And a little later, he added, "I should never make an end, if I attempted even to enumerate all the hidden changes of the Caterpillar and Worm kinds into winged insects; so fruitful and manifold is Nature in her genera and species!"[72]

[67] Réaumur criticizes Swammerdam for including in this class the insects that pupate inside other things, like plants or other insects' larvae; Réaumur, *Mémoires*, 1: 35–36.

[68] Swammerdam, *Book of nature*, 1: 119.

[69] Swammerdam, *Book of nature*, 1: 18.

[70] Jan Swammerdam, *Bybel der natuure.of Historie der insecten tot zeekere zoorten gebracht* (Leyden: by I. Severinus, 1737–38), 1: 39.

[71] Swammerdam, *Book of nature*, 2: 2.

[72] Swammerdam, *Book of nature*, 2: 100, 101.

Swammerdam had a large insect collection. He himself proclaimed, "I have in my musaeum specimens of all the insects I have spoken of in the foregoing treatise, to the number of more than twelve hundred, with their Nymphs, Chrysalides, and Vermiform-nymphs.... But I would not have the readers imagine, that I have minutely examined every single insect of my collection; I have only reduced a great many of them to my four orders." Indeed, Swammerdam echoed 16th-century botanists and adumbrated Réaumur, writing, "Let us do as much as we can, I believe it so far from possible to know every species of insects, and the changes of each, that I believe the labour of ages would not be sufficient to discover all the kinds of them." Nonetheless, he trusted in his method: "I challenge any one to produce a single insect, that may not be referred to one of the four orders of mutations which I have proposed; for to me, if I may here trust my reason, the thing appears impossible."[73] Was he "puffed up with pride" when he said that? If so, his successors would have said that the pride was justified.

Swammerdam's classification was enthusiastically adopted by the English naturalist John Ray. Ray's initial study of insects was conducted alongside Francis Willughby, and in 1678, Ray wrote to Aubrey that he thought Willughby's revised classification of insects for the never-published Latin edition of John Wilkins's *Essay* was the best that had been done.[74] When he initially returned to the serious study of insects in 1690, he proposed a classification along those lines, dividing insects into two categories: "those that do not undergo a change, and those that do suffer some kind of metamorphosis." The latter, in turn, were divided according to the number of feet their larvae possessed: many, six, or none.[75]

This classification collapsed larvae and adult forms into one group. It is unclear whether Ray drew this idea from Swammerdam or from Lister's Goedart. In his 1691 physico-theological work, *The Wisdom of God Manifested in the Works of Creation*, Ray justified his decision on pragmatic grounds. There were hundreds of different butterflies and beetles, "and if we should make the Caterpillers and Hexapods from whence these come to be distinct Species, as most Naturalists have done, the number will be doubled. But if those be admitted for distinct Species, I see no reason but their Aureliae also may pretend to a Specifick difference from the Caterpillers and Butterflies; and so we shall have 300 Species more, therefore we exclude both these from the degree of Species, making them to be the same Insect under a different larva or Habit."[76] While rendering

[73] Swammerdam, *Book of nature*, 2: 134–135.

[74] Ray, *Further Correspondence*, 165.

[75] Ray to Edward Lhwyd, July 7, 1690, in Ray, *Further Correspondence*, 208–209.

[76] John Ray, *The Wisdom of God Manifested in the Works of the Creation* (London: Printed for Samuel Smith, at the Princes Arms in St. Pauls Church-Yard, 1691), 6.

classification easier, Ray also underscored that change was a basic part of the nature of insects; one species might possess several distinct forms.

By 1692, Ray had discovered Swammerdam, of whom he spoke with the passion of a convert. Writing to his friend Edward Lhwyd, Ray pronounced Swammerdam's *Historia insectorum generalis* (which he read in the French translation) as "certainly ye best book that ever was written upon that subject."[77] Ray repeated his praise in nearly the same words in later letters to Hans Sloane and William Derham.[78] In 1703, as he turned his mind again to how he should classify insects, he decided that Swammerdam's method was best for insects that had both feet and (at some point in their lives) wings; sometimes he wrote as if he would use Swammerdam for his entire classification.[79]

The *Methodus insectorum* (*Method of Insects*) that Ray wrote in 1704, and that was published posthumously in 1705, shows Ray's debt to Swammerdam.[80] Ray retained Willughby's classification for insects that did not undergo metamorphosis. But now he divided the remaining ones into broad categories according to their modes of transformation, using Swammerdam's method, which he found the most pleasing.[81] Ray rejected Swammerdam's distinction between insects that metamorphosed outside and those whose metamorphosis took place within cases they made or within the bodies of other creatures, on the grounds that it was the nature of the metamorphosis, not its location, that mattered.[82] And he divided each group into smaller taxa, unlike Swammerdam, who was interested in broad patterns. Nonetheless, Ray's *Methodus*, and the posthumous *Historia insectorum* (*History of Insects*, 1710) that reprised it, cemented the idea, adumbrated by Goedaert and insisted upon by Swammerdam, that an insect species included all of the creature's metamorphoses from egg to adult.[83]

Swammerdam's classificatory scheme could prove problematic for certain practical purposes, however: in particular, for identification of an unknown insect. Presented with the creature as it existed at a particular moment in time, the naturalist might have no idea what kind of transformation, if any, it had undergone or would undergo in the future. This pragmatic problem motivated

[77] Ray to Edward Lhwyd, April 5, 1692, in Ray, *Further Correspondence*, 226–228.

[78] Ray to Hans Sloane, April 15, 1699, and to William Derham, late April or early May 1702, in Ray, *Correspondence*, 363–364, 399–400.

[79] Ray to Hans Sloane, February 23, 1702/3, and Ray to James Petiver, September 11, 1703, in Ray, *Correspondence*, 410–411, 434–435.

[80] John Ray, *Methodus insectorum, seu Insecta in methodum aliqualem digesta* (Londini: Apud Sam. Smith & Benj. Walford, Societatis Regiae Typographos, ad Insignia Principis in Area Boreali D. Pauli, 1705).

[81] Ray, *Methodus insectorum*, 6.

[82] Ray, *Methodus insectorum*, 12, 15.

[83] John Ray, *Historia insectorum: opus posthumum jussu Regia Societatis Londinensis editum; cui subjungitur appendix De scarabais Britannicis, autore M. Lister* (Londini: Impensis A. & J. Churchill, 1710).

Antonio Vallisnieri (1661–1730) to develop an alternate, ecological basis for classifying insects.[84]

Vallisnieri's classification appeared in his 1713 *Esperienze ed osservazione intorno all'origine, sviluppi, e costumi di varj insetti* (*Experiments and Observations on the Origin, Development, and Habits of Various Insects*).[85] The work's third tractate is devoted to "A new idea for the general division of insects." Vallisnieri's pragmatic orientation is signaled by the tractate's beginning, which notes how much damage insects can cause to garden plants; hence he is going to address how the insects themselves can be harmed. First, however, one must identify the insect that is causing damage, and that is difficult due to the enormous number of species. Vallisnieri had found more than 200 that nourish themselves from different parts of the oak tree alone.[86] Therefore, they must be classified in a way that permits their quick identification.

For this purpose, Vallisnieri proposed four general classes of insects: (1) those that live in plants and devour them; (2) those that grow and live in fluids, natural or artificial; (3) those that live on rocks, shells, and other solid objects; and (4) those that live on or within the bodies of animals, from humans all the way down to other insects, living or dead.[87] This division is analogous to the classification of diseases according to the part of the body that they afflict; to pursue the classification further, one would divide the first class into the different parts of the plant that they affect. "Such a series would require a thick book, but on the other hand it would, without confusion, quickly place before the eyes the insect we are seeking."[88]

In 1734, Réaumur would ridicule Vallisnieri's classification, not only because it returned to the older ecological criteria that had placed cetaceans and fish, or birds and bats, in the same class, but also because many insects spend different phases of their lives in different environments.[89] But Vallisnieri was not so thoughtless. "Once this first work of ordering is done," he continued, "we have another task ahead of us, just like the botanists: that is, to consider their final development, like considering the flower and seed of a plant." This will resolve the problem that many different insects live on the same parts of a plant: Considering both the place and the "final development" will allow the observer to distinguish the species.[90]

[84] On Vallisnieri (sometimes spelled Vallisneri), see Dario Generali, ed., *Antonio Vallisneri: La figura, il contesto, le immagine* (Florence: Leo S. Olschki, 2008).

[85] I have consulted the second edition: Antonio Vallisnieri, *Esperienze ed osservazione intorno all'origine, sviluppi, e costumi di varj insetti, con atre spettanti alla naturale, e medica storia* (Padova: Nella Stamperia del Seminario, appresso Gio: Manfrè, 1726).

[86] Vallisnieri, *Esperienze ed osservazione*, 40.

[87] Vallisnieri, *Esperienze ed osservazione*, 42–43.

[88] Vallisnieri, *Esperienze ed osservazione*, 43–44.

[89] Réaumur, *Mémoires*, 1: 37–38.

[90] Vallisnieri, *Esperienze ed osservazione*, 44.

Even Vallisnieri, then, accepted Swammerdam's principle. His classification based on final development involved a firm rejection of the principle that worms, nymphs, or chrysalises could be the basis of classification. Instead, he divided insects into those that have wings in the final form and those that do not; the former, in turn, are divided by the number and nature of the wings. Turning to wingless insects, Vallisnieri reminded his readers of his starting principle: The embryonic forms of winged insects do not belong in this group, but rather with the adult winged insects they will become.[91] Vallisnieri may have thought that metamorphosis was not the best starting place for a pragmatic classification, but he recognized that it was a fundamental aspect of what it meant to be an insect species.

Ray and Vallisnieri were not alone in accepting Swammerdam's authority and principles. The Nuremberg miniaturist and insect-lover August Johann Rösel von Rosenhof attempted to join Ray's extensive catalogue of insect species with Swammerdam's desire for a detailed description of transformation by publishing his *Monatlich herausgegebene Insecten-Belustigung* (*Monthly Insect Entertainment*), a series that every month, for nearly two decades, gave its purchasers a new insect, with its habits and transformations carefully described in eight quarto pages and its morphology, from egg through larval instars and pupal stages to imago, depicted in hand-colored engravings.[92] Rather than pursue Swammerdam's influence and echoes to the present, however, I would like to end in the early decades of the 18th century, with the reflections of René-Antoine Ferchauld de Réaumur on the problems and limits of insect classification.

Réaumur, admitted to the Académie des Sciences as an *élève géometre*, had polymathic interests.[93] He wrote a multivolume set of *Mémoires pour servir à l'histoire des insectes* (*Memoirs to Serve for the History of Insects*), published in the 1730s and '40s. As Réaumur's choice of title indicates, he was diffident about the possibility of writing a history of insects. Not enough facts had yet been gathered, he wrote, and many of those that had been published were dubious. Like Swammerdam, he singled out Goedaert as an author who deserves praise

[91] Vallisnieri, *Esperienze ed osservazione*, 66–70.

[92] August Johann Rösel von Rosenhof, *Der monatlich herausgegebenen Insecten-Belustigung erster[-vierter] Theil*, 4 vols. (Nürnberg: gedruckt bey Johann Joseph Fleischmann, 1746–61). On Rösel, see Brian W. Ogilvie, "The Pleasure of Describing: Art and Science in August Johann Rösel von Rosenhof's *Monthly Insect Entertainment*," in *Animals on Display: The Creaturely in Museums, Zoos, and Natural History*, eds. Liv Emma Thorsen, Karen A. Rader, and Adam Dodd (University Park: Penn State University Press, forthcoming).

[93] See Jean Torlais, *Un esprit encyclopédique en dehors de l'Encyclopédie: Réaumur, d'après des documents inédits* (Paris: Librairie scientifique et technique Albert Blanchard, 1961); Gilles Bresson, *Réaumur: Le savant qui osa croiser une poule avec un lapin* (Le Château d'Olonne: D'Orbestier, 2001); Yves Carton, "Réaumur (1683–1757): The Discoverer of Parasitoids in France," *Biological Control* 32.1 (2005): 40–48; and Mary Terrall's forthcoming study of Réaumur's natural history.

tempered with blame; though a careful observer in many regards, he often let his imagination take advantage of his reason.

An equally great problem, though, was the sheer number of insects. Réaumur was obsessed with "the prodigious variety of forms of insects of different classes and genera."[94] Every country has its own insects, and every year, each country produces more. Botanists know how many varieties of plants there are; think, Réaumur told his reader, how many of those plants support insects that live only on them—sometimes more than one species?

What was needed, Réaumur thought, was certainly not more descriptions of insect species. Rather, we needed more *facts* about insects: knowledge about their lives, and above all, about their behavior (1: 13) that would help us make sense of their immense variety. "It will suffice for us," he wrote, "and our curiosity ought to be content to know the principle genera, and especially to know those that we see often... I admit that I would be left unmoved by an enumeration of all the species of each genera, even if we could do it; it is enough to consider those that have shown us that they deserve our consideration."[95] Much of the pleasure in seeing insects comes from knowing what they were before, or what they will become; therefore, Réaumur argued, it is best to concentrate on those insects that we see often, "those with whom we live, so to speak."[96]

What Réaumur proposed as a solution to insect diversity is an anthropocentric ordering of insects. It was not arbitrary, since it rested on insects' behavior, their anatomy, or "other striking aspects" of them. However, "some small differences in the forms of wings or legs, or the variety of colors" are not grounds for bothering to make careful distinctions; in such cases "we can leave them confounded with one another."[97] The anthropocentric element was a principle of selection: We should study those insects that live with us, that enter into our daily lives, and those insects as well that teach us something new about what it means to be an insect.

Réaumur was not indifferent to classification. He singled out Swammerdam for particular praise and gave a synopsis of his system taken from the *Historia insectorum generalis*. Like Ray, he criticized details of Swammerdam's scheme: He announced that he would show, for instance, that the mayfly does not properly fit in the second of Swammerdam's classes, where Swammerdam placed it, or in the third. Like Ray, he also criticized Swammerdam for including in the fourth class those insects that pupate within plants, other insects, etc., since they properly belong to the third class.[98]

[94] Réaumur, *Mémoires*, 1: 14.

[95] Réaumur, *Mémoires*, 1: 3.

[96] Réaumur, *Mémoires*, 1: 17–18.

[97] Réaumur, *Mémoires*, 1:3.

[98] Réaumur, *Mémoires*, 1: 33, 35–36.

Swammerdam's classification had other faults too. "Its major drawback is that uses too few divisions. Four classes are not enough to allow distinguishing such an innumerable quantity of insect genera that have such differences between them."[99] Vallisnieri's classification fails because it does not account for the fact that many insects live in different places at different points in their lives; by focusing on habitat, it neglects transformation. (As we have seen, this is not an entirely fair critique.) Réaumur had imbibed Goedaert's and Swammerdam's principle: Metamorphosis is fundamental, and a classification that does not treat the insect diachronically is no classification at all. Yet metamorphosis by itself, as Ray realized, was not enough.

Réaumur set aside the further task of producing a complete classification. "After all," he concluded reassuringly—or perhaps to reassure himself—

> it is perhaps not as important to put the facts in a good order than to assemble enough that deserve attention. Any order is perhaps good, as long as one gives continuous histories of each principal species of insect. But we must unequivocally reject the order in which Aldrovandi, Moffett, and Johnston treated caterpillars and butterflies. In one chapter they put all the butterflies that they knew, and in another, all the caterpillars that they knew. It would have been better to have a smaller series of caterpillars and butterflies, and to find each caterpillar associated with its butterfly. As soon as we have a history of insects with enough facts, no matter how it is organized, it will be easy to reduce it to the method of Swammerdam, or that of Vallisnieri, or any other: all that will be necessary is to draw up Tables in which the insects are placed in the order one wants.[100]

In short, Réaumur approved of the observational work on species and metamorphosis that had been pursued by Goedaert, continued by Swammerdam and Ray, and that Rösel and others would continue to pursue throughout the 18th century—not least Réaumur himself. Classification, he thought, would follow naturally.

5. Conclusion: Insect Order before Linnaeus

Was Réaumur too optimistic? At first glance, his confidence seems naïve, but I'm not so sure. Modern entomology courses—following a pattern established in

[99] Réaumur, *Mémoires*, 1: 36.
[100] Réaumur, *Mémoires*, 1: 38.

the late Renaissance for botany by Adriaan van de Spiegel—introduce students to the major insect orders, choosing within them "typical" families, genera, and species to instantiate the basic insect *Baupläne* and developmental histories.[101] Specialists in entomology learn more, but how much depends on the narrowness of their specialties. Taxonomists revising a genus or family will know it cold. But eminent entomologists and fellow travelers, like Karl von Frisch or Thomas Eisner, might know only a few hundred or a few thousand species—nothing at all in the vast sea of insect diversity.

Perhaps a quarter of all known animal species are Coleoptera. Over half are insects; even more are arthropods.[102] For 17th-century naturalists, whose notion of "insect" was even more expansive, but whose enumeration of species was much smaller than ours, insects comprised an enormous proportion of animals. In 1691, John Ray estimated that there were perhaps 225 species of quadrupeds, 750 species of birds, and 1,000 species of fish, while insects (aquatic and terrestrial) comprised around 10,000 species—over five-sixths of the total. A year later, he doubled his estimate for insects, which now comprised nearly 90 percent of animal species.[103] Sustained attention to insect diversity resulted in many more species being identified and named.

But in a certain sense, there were fewer species: Caterpillars joined butterflies, grubs joined beetles as successive stages of the same species; they were not distinct species with their own, separate life histories. Despite a few rearguard attempts to preserve it, the ancient and Renaissance notion of metamorphosis, the idea that when a caterpillar formed a chrysalis, it had ended its life and a new species would emerge as a butterfly, had run its course. Swammerdam won: Slow, continuous changes, following ironclad rules—often including the rules of parasitism—were the hallmark of Enlightened insectology.

And yet a sense of wonder remained. Swammerdam, Merian, Rösel, and Réaumur loved insects. Many were beautiful; even those that were not followed a beautiful pattern of change that was imbued in them by the Creator. Studying that change was a way to come closer to God. Overwhelmed by insects' sheer numbers and diversity, these early modern naturalists, like Gass's anonymous narrator, succeeded—at least for a time—in finding an order of insects.

[101] Adriaan van de Spiegel, *Isagoges in rem herbariam libri duo* (Patavii: Apud Paulum Meiettum, ex typographia Laurentii Pasquati, 1606); David Grimaldi and Michael S. Engel, *Evolution of the Insects* (Cambridge, UK: Cambridge University Press, 2005). A Bauplan (plural Baupläne) is the set of morphological features found in all species within a given biological taxon.

[102] Estimates of the total number of insect species, known and unknown, vary widely. Grimaldi and Engel, *Evolution*, 11-15.

[103] Ray, *Wisdom* (1st ed., 1691), 5–7; John Ray, *The Wisdom of God Manifested in the Works of the Creation* (London: Printed for Samuel Smith, at the Princes Arms in St. Paul's Church-yard, 1692), 6–9.

NAME INDEX

SUBJECT INDEX